Characteristic classes of foliations

H V Pittie

Courant Institute of Mathematical Sciences

Characteristic classes of foliations

Pitman Publishing

LONDON · SAN FRANCISCO · MELBOURNE

AMS Subject Classifications: (main) 57D30, 57D20
(subsidiary) 17B55, 18H25, 53C05

PITMAN PUBLISHING LTD
Pitman House, 39 Parker Street, London WC2B 5PB, UK

PITMAN PUBLISHING CORPORATION
6 Davis Drive, Belmont, California 94002, USA

PITMAN PUBLISHING PTY LTD
Pitman House, 158 Bouverie Street, Carlton, Victoria 3053, Australia

PITMAN PUBLISHING
COPP CLARK PUBLISHING
517 Wellington Street West, Toronto M5V 1G1, Canada

SIR ISAAC PITMAN AND SONS LTD
Banda Street, PO Box 46038, Nairobi, Kenya

PITMAN PUBLISHING CO SA (PTY) LTD
Craighall Mews, Jan Smuts Avenue, Craighall Park,
Johannesburg 2001, South Africa

Library of Congress Cataloging in Publication Data
Pittie, H V 1944–

Characteristic classes of foliations.

(Research notes in mathematics; 10)
1. Foliations (Mathematics) 2. Characteristic
classes. I. Title. II. Series.
QA613.62.P57 514′.7 76–26499
ISBN 0–273–00311–9

ISBN 0 273 00311 9

Reproduced and printed by photolithography and bound in
Great Britain at The Pitman Press, Bath

3512:97

Preface

These notes are a revised version of lectures notes from Warwick, which I wrote up from my talks there in the spring of 1975. They are intended as an exposition of the main constructions and results in the theory of secondary characteristic classes of foliations, except that they do not bring in Haefliger's classifying space. My aim was not to give an encyclopedic treatment of the theory of secondary classes, but rather to give a couple of swift and computable ways of defining them, and to work out several examples. One purpose which I hope these notes will serve well is to provide proofs and computations (my own) for many examples which have been announced in the literature but whose proofs have not yet seen the light of day: in a couple of instances there were errors in the announced calculations which I have pointed out in one instance and corrected in the other.

My own interest in these classes arose from the fact that they are susceptible of continuous variation or deformation (in parameters), unlike the ordinary (primary) characteristic classes of bundles, and so I took it upon myself to give a simplified account in English of the paper of Gelfand, Feigin and Fuks [8] in Chapter IV. The appendices of course contain standard material, but the proof of the Chevalley-Eilenberg theorem (in Appendix II) is simpler than any proof in the literature that I have seen.

My main sources for these notes were Bott's Mexico lectures [2], the papers of Gelfand, Fuks et al., Haefliger's and Godbillon's Bourbaki exposés and some lecture notes of Atiyah from Canberra (1972).

It remains to express my sincere thanks to George Lusztig who invited me to spend the year at Warwick, and was a constant source of helpful and stimulating remarks: to Robert V. Kohn for having waded through the original hand-written manuscript, suggesting improvements and catching errors: and of course to Connie Engle for her usual expert and efficient typing.

Harsh Pittie

TABLE OF CONTENTS

Contents

I Preliminary results

1. Bott's Vanishing Theorem

We shall be concerned entirely with smooth (i.e. C^∞), paracompact manifolds M (without boundary, but not necessarily compact) and with smooth foliations on them. In fact for our purposes, a smooth foliation on M will be a smooth subbundle $E \to M$ of constant rank of the tangent bundle TM which is integrable in the sense that if X, Y are vector-fields in E, their bracket [X,Y] is also in E. By the classical theorem of Frobenius (and apparently of Clebsch and Deahna) this is equivalent to the more geometric definition of a foliation as a decomposition of M into disjoint, smooth submanifolds of fixed dimension, smoothly fitted together. Recent results of Thurston [20] show that smoothness is a genuine restriction in the theory of foliations, so that one does not have the general blessing of Whitney under which differential topologists have worked for so long. However, at least for the time being the C^∞-setting seems to be the proper one for the theory of (secondary) characteristic classes.

The cohomology of M, written $H^*(M)$, will refer to the real cohomology, generally regarded as the de Rham cohomology. On occasion the complex cohomology will slip in, but this will always be specified. Thus the characteristic classes we discuss will be real (or complex) cohomology classes. For a real vector-bundle $V \to M$, $\mathrm{Pont}^*(V)$ will be the graded subring of $H^*(M)$

generated by the Pontryagin classes; similarly for a complex vector-bundle, $\text{Chern}^*(V)$ will denote the graded subring generated by the Chern classes. Note that $\text{Chern}^*(V) \subseteq H^*(M)$; the complex cohomology classes which crop up later on are really certain complex linear combinations of the Chern classes which are better-suited to our calculations.

With this understood, we can state the basic result of Bott [2] which underlies the whole development of secondary characteristic classes of foliations.

Theorem 1 (Bott [2]). Let $E \subseteq TM$ be an integrable subbundle of codimension q. Then $\text{Pont}^j(TM/E) = 0$ for $j > 2q$.

Proof. The proof is an immediate consequence of two facts; first, the Chern-Weil theorem which says that the Pontryagin ring of a bundle may be calculated from the curvature of any connection, and second, that for E as above the normal bundle $Q = TM/E$ has a particular connection (variously called basic, torsion-free, etc.) whose curvature tensor raised to powers $> q$ is zero. As the curvature tensor is locally a matrix of 2-forms, this says that the Pontryagin forms of this curvature will be zero in degrees greater than 2q. We shall now prove the second fact -- actually for the dual bundle; this is essentially the same thing and gives a better computational grip for later use. For the second assertion as stated, see Bott [2].

So consider $Q = TM/E$, and its dual $Q^* \subseteq TM^*$, which is the bundle of all cotangent vectors annihilating E. In terms of Q^*, the integrability of E can be phrased thus: for every set of q linearly independent 1-forms $\omega_1,\ldots,\omega_q$ which generate $\Gamma(Q^*|_U)$, there are 1-forms $\theta_{ij} \in A_U^1(M)$ such that

$$d\omega_i = \sum_{j=1}^{q} \theta_{ij} \wedge \omega_j , \qquad 1 \leq i \leq q .$$

More globally this says that the global sections $\Gamma(Q^*) \subseteq A^1(M)$ generate a <u>differential</u> ideal in the algebra $A^*(M)$ of all smooth forms. Now over U, we can define a connection D_U on $Q^*|_U$ by assigning to $\omega_1,\ldots,\omega_q$ any q sections $s_1,\ldots,s_q \in \Gamma(QM^* \times Q^*|_U)$ and extending D_U to $\Gamma(Q^*|_U)$ by the requirements that (i) D_U be additive, (ii) for any smooth $f\colon U \to R$, $D_U f\omega_i = fD_U\omega_i + df \otimes \omega_i$. We choose to do this by setting

$$D_U\omega_i = \sum_{j=1}^{q} \theta_{ij} \otimes \omega_j$$

for some θ_{ij} as above. Then we can get a global connection by patching these local connections together by a partition of unity. The virtue of our choice is that over any open set $U' \subset M$ for which there are generating sections $\omega_1',\ldots,\omega_q'$ for $\Gamma(Q^*|_{U'})$, the image of $D_U\omega_i'$ in $A^2(M)$ coincides with $d\omega_i'$; this follows easily from the fact that d satisfies (i) and (ii) as well. In particular, given a local basis, we can read off the local connection-form -- i.e. the matrix (θ_{ij}) -- by differentiating the local basis forms.

It is also clear that we then get some relation on the curvature of such a connection, by differentiating the identity for $d\omega_i$. To wit, if over U we have

$$d\omega_i = \sum \theta_{ij} \wedge \omega_j$$

then applying d, and eliminating $d\omega_j$ we get

$$0 = \sum_j d\theta_{ij} \wedge \omega_j - \sum_j \theta_{ij} \wedge \left(\sum_k \theta_{jk} \wedge \omega_k \right)$$

$$0 = \sum_k \left(d\theta_{ik} - \sum_j \theta_{ij} \wedge \theta_{jk}\right) \wedge \omega_k \ .$$

Now $K_{ik} = d\theta_{ik} - \sum_j \theta_{ij} \wedge \theta_{jk}$ is simply the (i,k) entry in the curvature matrix for D (over U) and our relation reads

$$\sum_k K_{ik} \wedge \omega_k = 0 \ .$$

If we choose vector-fields $Z_1, \ldots, Z_q$ dual to the ω_i , and let $X, Y \in \Gamma(E|_U)$ then applying (X, Y, Z_ℓ) to the last equation above, we find that $K_{i\ell}(X,Y) = 0$. Hence each $K_{i\ell}$ is in the ideal generated by $\omega_1, \ldots, \omega_q$. In particular, if ϕ is an invariant polynomial of degree > q, then $\phi((K_{i\ell})) = 0$, as was to be shown. □

There is a complex-analytic version of this theorem as well, proved in exactly the same way. The analogous connection there is automatically of type (1,0).

Theorem 1. Let M be a complex manifold, TM' its complex tangent bundle, and $E' \subseteq TM'$ an integrable subbundle of codim q. Then

$$\text{Chern}^j(TM'/E') = 0 \quad \text{for} \quad j > 2q.$$

Bott [2] used these theorems to show that there is a topological obstruction to integrability, i.e. not every subbundle of TM is homotopic (even abstractly) to an integrable one. A further refinement concerning Massey products has been given by H. Shulman [18].

A topologist is naturally led to ask whether Theorems 1 and 1' hold for the integral characteristic classes as well, and whether they admit a "topological" proof. The remarks below indicate negative answers to both questions.

Bott and Heitsch [4] have given simple examples to show that the <u>integral</u> Pontryagin ring need not vanish above twice the codimension. Their construction runs as follows. Consider the group $\mu_n = \{\lambda \in C: \lambda^n = 1\}$. It acts on R^2 by rotations of multiples of $2\pi/n$, and on the odd-sphere $S^{2k-1} = \{(z_1,\ldots,z_k) \in C^k: \{|z_1|^2 = 1\}$ by $\lambda(z_1,\ldots,z_k) = (\lambda z_1,\ldots,\lambda z_k)$. Thus it also acts on the product $S^{2k-1} \times R^2$, and freely, since the action is free on the first factor. Let $M = (S^{2k-1} \times R^2)/\mu_n$. Now on $S^{2k-1} \times R^2$ there is a 2-form $\tilde{\omega}$ which restricts on each plane $\{z\} \times R^2$ to the volume-form $dx \wedge dy$. Since it is clearly invariant under μ_n, $\tilde{\omega}$ descends to a nonvanishing 2-form ω on M which still satisfies $d\omega = 0$. Hence ω defines a codimension 2 foliation with oriented normal bundle; indeed, M is the "standard" (flat) oriented 2-plane bundle over the lens space $X = S^{2k-1}/\mu_{n_1}$, and the foliation is the one transverse to the fibres, so the normal bundle defined by ω is precisely $\pi^*(M)$, where $\pi: M \to X$ is the bundle projection.

Now $p_1(\pi^*(M)) = \pi^*(p_1(M))$ and π is a homotopy-equivalence, so we may as well show that $p_1(M)^2 \in H^8(X:Z)$ is not zero. So take $k \geq 5$: since the even-cohomology of X is given by $H^{ev}(X:Z) = (Z/n)[x]/x^k$, $\deg x = 2$, and $p_1(M) = x^2$, we are done.

One can easily modify this example so that the foliation is on a compact manifold: simply replace R^2 by S^2 with μ_n

rotating the sphere around the north-south axis. The quotient by μ_n is then an S^2-bundle over X, and the sections at the north and south poles are leaves. The normal bundle is still the same R^2-bundle, and it suffices to show that $p_1(M)^2 \neq 0$ when restricted to the leaf through the north (or south) pole.

Recently Thurston has proved a spate of remarkable existence theorems about foliations on compact manifolds. His results show that on a compact manifold, any subbundle of the tangent bundle of codimension > 1 (and also codim = 1 if everything is oriented)* is homotopic to a C^0-integrable subbundle with smooth leaves. Now it is easy to find examples of subbundles whose Pontryagin ring does not vanish above twice the codimension, even in the real cohomology: for example, one could take Bott's codimension 2 example on CP_{2k+1} ([2], pp. 35, 36). Thus Bott's vanishing theorem depends essentially on the C^2-differentiability; and whatever else one might mean by a "topological" proof, it ought to be one that doesn't require such restrictions.

2. The Godbillon-Vey class

We shall now show how Bott's vanishing theorem enables one to construct secondary or exotic classes for foliations, even in the simple case of a codimension-one foliation. This class was first constructed by Godbillon and Vey [10], independently of Bott's theorem, and their methods can also be pushed to yield further invariants in higher codimension (see Chapter 3): for the moment we shall concentrate on using Bott's result.

---

*In [20], Theorem 2 only states this for codimension > 1; a note added in the proof completes the theorems to codim 1.

So consider an integrable codimension-one subbundle $E \subseteq TM$; we assume solely for convenience that the normal bundle $Q = TM/E$ is trivial. Then E is specified by a nonvanishing 1-form $\omega\colon TM \to R$ which satisfies the integrability condition $d\omega = \theta \wedge \omega$ for some $\theta \in A^1(M)$.

Theorem 2. $\theta \wedge d\theta$ is a closed form. Its cohomology class $\{\theta \wedge d\theta\} \in H^3(M{:}R)$ is independent of all choices, and an invariant of the transverse or "integrable" homotopy-class of the foliation.

Proof. First recall from the proof of Theorem 1 that θ must be the connection-form of some torsion-free connection on Q^*. Hence the curvature of the connection is given by $d\theta$, and so by Theorem 1 again $d\theta \wedge d\theta = 0$. Now $d(\theta \wedge d\theta) = d\theta \wedge d\theta = 0$, so $\theta \wedge d\theta$ is closed.

Of course, θ and ω are not unique; ω may be replaced by any form $f\omega$, where $f\colon M \to R^*$ is smooth, and θ may be replaced by $\theta + g\omega$, where $g\colon M \to R$ is smooth. First consider the effect of replacing ω by $f\omega$: we have

$$d(f\omega) = f\,d\omega + df \wedge \omega = f\theta \wedge \omega + df \wedge \omega = (\theta + \frac{df}{f}) \wedge f\omega .$$

If we set $\xi = \theta + df/f$ and note that df/f is closed* then $\xi \wedge d\xi = \theta \wedge d\theta + df/f \wedge d\theta$, and the second term is exact. Now consider the effect of replacing θ by $\eta = \theta + g\omega$ (and recall that by differentiating $d\omega = \theta \wedge \omega$ we get $d\theta \wedge \omega = 0$). Then

---

*Of course, in the real case it is exact; our proof is given so it adapts without change to Theorem 2' below.

$$\eta \wedge d\eta = (\theta + g\omega) \wedge (d\theta + g\, d\omega + dg \wedge \omega) = \theta \wedge d\theta - dg \wedge \theta \wedge \omega$$
$$= \theta \wedge d\theta - dg \wedge d\omega$$

so again the second term is exact. The cohomology class $\{\theta \wedge d\theta\}$ is independent of the choices involved in picking ω and θ.

In fact these explicit calculations could have been dispensed with by showing first -- as we shall now do -- that if $E \subset T(M \times [0,1])$ is an integrable subbundle of codim 1, such that for each t, $E_t \subset T(M \times \{t\})$ gives a codimension one integrable subbundle, then $\{\theta_{E_0} \wedge d\theta_{E_0}\} = \{\theta_{E_1} \wedge d\theta_{E_1}\}$. One finds that ω: $T(M \times [0,1]) \to R$ restricts locally to ω_t giving E_t, and similarly that there is a global θ such that $d\omega_t = \theta_t \wedge \omega_t$. Then $\{\theta \wedge d\theta\} \in H^3(M \times [0,1])$ restricts at one end to $\{\theta_0 \wedge d\theta_0\}$ and at the other end to $\{\theta_1 \wedge d\theta_1\}$, and so by homotopy invariance they are the same class. □

Now if M is a complex manifold, $E' \subset TM'$ an integrable subbundle of the complex tangent bundle of codimension 1 -- with TM'/E' trivial (no longer simply a convenient assumption) -- then again one can define $\theta' \wedge d\theta' \in H^3(M:C)$ and we have

<u>Theorem 2'</u>. $\{\theta' \wedge d\theta'\} \in H^3(M:C)$ is a well defined invariant of the integrable homotopy class of E.

We now mention some trivial examples for which the Godbillon-Vey class is zero, and attempt to give a geometric interpretation to what it measures.

<u>Remarks</u>. 1. Let L stand for R or S^1, and α for the "standard" 1-form on L, -- that is dt or "$d\theta$". If f: $M \to L$ is the projec-

tion of a fibre bundle, then M is foliated by the fibres of f, and for this case the form dual to the integrable subbundle is precisely $\omega = f^*(\alpha)$. But $d\omega = 0$, so clearly the Godbillon-Vey class is zero as well.

In a similar vein, if V is a manifold, $f: V \to R$ a smooth function with 0 as a regular value, then $V \times L$ has a codimension one foliation; its 1-form is $\omega = f\alpha + df$ (which is nowhere zero by the hypothesis on f, and integrable since $d\omega = df \wedge \alpha = \omega \wedge \alpha$). Since $d\alpha = 0$, $\alpha \wedge d\alpha = 0$.

It is clear from the remarks above that the Godbillon-Vey class measures some sort of "twisting" of the leaves. To make this precise, put a Riemannian metric on TM, and let $X \in \Gamma(TM)$ be a vector-field of norm 1 in $E^{\perp}$ -- i.e., normal to the leaves. Take ω to be the 1-form dual to X -- i.e., ω annihilates E and satisfies $\omega(X) = 1$ at all points of M. Then we can express the Godbillon-Vey class in terms of ω and its Lie derivatives along X, as follows. First, $L_X\omega = i_X d\omega + di_X\omega = i_X\, d\omega$, and so $d\omega = (L_X\omega) \wedge \omega$. Differentiating this we find that $(dL_X\omega) \wedge \omega = 0$, or $L_X^2\omega \wedge \omega = dL_X\omega$. The Godbillon-Vey class is $\{L_X\omega \wedge dL_X\omega\}$ or what is the same by the last equation, $\{\omega \wedge L_X\omega \wedge L_X^2\omega\}$.

Now consider what these quantities are geometrically. $L_X\omega$ is the derivative of ω along the normal to the leaves, and $L_X^2\omega$ is the normal derivative of $L_X\omega$. Thus $\omega \wedge L_X\omega$ measures the tendency of a leaf to turn away from the previous nearby ones, and $dL_X\omega = L_X(L_X\omega \wedge \omega)$ is the velocity in the normal direction with which the leaves are turning away. Hence the form $\omega \wedge L_X\omega \wedge L_X^2\omega$ measures something like a gyroscopic precession.

Thurston [21] has given this geometric phenomenon the inspired name of "helical wobble"; see his paper [21] for a very convincing picture.

3. The examples of Roussarie, Thurston and Bott

We now give examples to show that the Godbillon-Vey class can be nonzero. In fact Thurston's examples show that it can take on any real value and Bott's show that the complex G-V class can take on any complex value.

Roussarie's examples. Consider the 3-dimensional Lie group $SL(2,R)$ of real 2×2 unimodular matrices and within it a discrete subgroup Γ such that the quotient $SL(2,R)/\Gamma$ is compact: such subgroups exist in abundance as we shall show below. On $M = SL(2,R)/\Gamma$ we shall construct a codimension 1 foliation whose Godbillon-Vey invariant in $H^3(M:R)$ is not zero.

To construct the foliation, it suffices to exhibit a non-vanishing 1-form ω on M such that $\omega \wedge d\omega = 0$. Now any left invariant 1-form on $SL(2,R)$ descends to M, so it suffices to give one on $SL(2,R)$: in fact the whole computation will be with left invariant forms, so we shall work on the group itself. $SL(2,R)$ has 3 linearly independent, left invariant 1-forms θ, ξ, η which satisfy the relations* $d\theta = -\xi \wedge \eta$, $d\xi = -2\theta\wedge\xi$, $d\eta = 2\theta \wedge \eta$. We choose η to define the foliation: then the G-V

---

*These are easily read off from the dual bracket relations $[H,X] = 2X$, $[X,Y] = H$ and $[H,Y] = -2Y$ satisfied by $H = \begin{pmatrix} 1 & 0 \\ 0 & -1 \end{pmatrix}$, $X = \begin{pmatrix} 0 & 1 \\ 0 & 0 \end{pmatrix}$, $Y = \begin{pmatrix} 0 & 0 \\ 1 & 0 \end{pmatrix}$.

form is $4\theta \wedge d\theta = -4\theta \wedge \xi \wedge \eta$. Now $\theta \wedge \xi \wedge \eta$ is a volume form on the group, hence also on M: and as M is compact, it is non-zero.

The leaves of the foliation can be described as follows. On the group itself, η annihilates the space spanned by H and X (see footnote) which is the Lie algebra of the subgroup $U = \left\{\begin{pmatrix} a & b \\ 0 & a^{-1}\end{pmatrix} : a > 0\right\}$. Thus on the group the form η defines the "trivial" foliation by cosets of U. On M therefore, the leaf through the point $g\Gamma$ is $g_*(U/U \cap g\Gamma g^{-1})$. In Chapter 2, in the fourth section, this foliation is described in terms of geodesics in the upper half plane.

Now for a special subclass of Γ's one can give a very explicit geometric interpretation of this G-V class. Suppose S is a compact Riemann surface of genus ≥ 2: then its universal cover can be identified with the upper half-plane H. On H, the group of holomorphic isomorphisms is given by SL(2,R) acting through Mobius transformations $z \to (az+b)/(dz+d)$. Hence $\pi_1(S)$, interpreted as a group of deck transformations, corresponds to a discrete subgroup π of SL(2,R) * which is an extension of $\pi_1(S)$ by $\underset{\sim}{Z}/2$. Since $H \simeq SL(2,R)/SO(2)$, we see that $\pi_1(S) \backslash H \simeq \pi \backslash SL(2,R)/SO(2) \simeq S$ is compact, hence $SL(2,R)/\pi$ is compact as well. This gives the construction of such discrete subgroups promised earlier. In particular, we see that when $\Gamma = \pi_1(S)$, M is the tangent circle-bundle over S. Moreover, the form $\theta = \frac{1}{2}$ line element along the circles, so

*PSL(2,R) = SL(2,R)/{±1}: so for our Lie algebra computations, there is no difference.

$$4\int_M \theta \wedge \xi \wedge \eta = 4\int_{SO(2)} \theta \int_S \xi \wedge \eta = 4\pi \int_S \xi \wedge \eta = 4\pi \quad \text{(area of S)}.$$

The normalizations of ξ and η have been so chosen that if we construct a regular (non-Euclidean) polygon $P_\Gamma \subset H$ mapping onto S, then $\int_{P_\Gamma} \xi \wedge \eta$ = non-Euclidean area of $P_\Gamma = (g-2)\pi - \sum \alpha_i$ where α_i = angles in P_Γ. This last computation is simply the Gauss-Bonnet theorem.

Thurston's examples. Starting from Roussarie's foliations, Thurston [21] constructs a family of codimension-one foliations on S^3 whose Godbillon-Vey class varies continuously. In order to understand the general construction, we begin by looking at a special subclass.

Fix a closed Riemann surface S of genus ≥ 2, represented as $SO(2) \backslash SL(2,R)/\Gamma$. Then the group of (holomorphic) automorphisms $A(S) \simeq N(\Gamma)/\Gamma$ where $N(\Gamma)$ = normalizer of Γ in $SL(2,R)$; it is a finite group. Pick a nonconstant meromorphic function $f: S \to P_1$; since f takes each value the same number of times (counted with multiplicity) f is a ramified covering. The group of covering transformations of f, C(f) is a subgroup of A(S). Let $z_1,\ldots,z_k \in P_1$ be the ramification points, $\{p_{ij}\} = f^{-1}(z_i)$: since each $\gamma \in C(f)$ takes the p_{ij} into themselves, it induces a covering map of $S - \{p_{ij}\} \xrightarrow{\bar{f}} P_1 - \{z_1,\ldots,z_k\}$. By an elementary completeness argument, C(f) is the group of covering transformations of this restricted (and regular) covering $\bar{f}$.

Now over P_1 we have the Hopf-bundle. $S^1 \to S^3 \to P_1$, and pulling this back via $\bar{f}$, we may identify the inverse image as the bundle $SO(2) \to SL(2,R)/\Gamma \to S$ restricted to $S - \{p_{ij}\}$. (More precisely, make the argument for the tangent circle bundle of P_1 which pulls back to the same creature over any covering, and pass to double covers.) Now the covering group of $\bar{f}$: $(SL(2,R)/\Gamma - \{SO(2)_{ij}\}) \to (S^3 - \{S_i^1\})$ is still $C(f) \subseteq N(\Gamma)/\Gamma$. As the foliation of Roussarie is invariant under the right action of $N(\Gamma)/\Gamma$, we can push the foliation onto $S^3 - \{S_{ij}^1\}$. It remains to close it up over all of S^3. For this purpose, one distends the original holes z_i in P_1 to small disks, so that the foliation now is defined over S^3 - {k solid tori}; and one fills in each solid-torus with a Reeb component (checking that on the boundary of the solid tori the foliation is given by lines of rational slope).

The Godbillon-Vey class here is easy to compute. In the complement of the exceptional circles, the foliation is pushed from a finite covering, so the form $\theta \wedge \xi \wedge \eta$ pushes down to $\frac{1}{n}(\theta \wedge \xi \wedge \eta)$ on $S^3 - \{S_i^1\}$, where n = degree of f. Modifying it in arbitrarily small tubular neighborhoods has no effect on the integral, so $\int_{S^3}$ G-V class $= \frac{1}{n}\int \theta \wedge \xi \wedge \eta = \frac{1}{n}(4\pi \text{ area } \Gamma)$. Performing this construction over all closed surfaces S, we can get any rational multiple of π^2 as a value for the Godbillon-Vey class.

To get the irrational values, one has to go back to Thurston's method. One starts with a (hyperbolically) convex polygon P in the upper half-plane or unit disk, which may not be regular -- i.e. may not be a fundamental domain for any closed surface S. Label the sides $S_1,\ldots,S_k$; reflect P through one of the sides, say S_1 to get P', and let $Q = P \cup P'$ be the polygon P-doubled. Remove miniscule discs around each vertex of Q, and let $\bar{Q}$ be the remaining portion of Q.

Now the tangent circle bundle of H (upper half-plane) is SL(2,R): let π: SL(2,R) $\to$ H be the projection. $\pi^{-1}(\bar{Q}) \simeq \bar{Q} \times SO(2)$; but there is a unique map $\gamma_i \in SL(2,R)$ identifying a side $s_i \in P$ with $s_i' \in P'$, and if we use these to make identifications in $\bar{Q} \times SO(2)$ we end up with a manifold $M = (S^2 - k \text{ punctures}) \times S' \simeq S^3 - \{k \text{ solid tori}\}$. (The way I think of this is to put P in the lower hemisphere of S^2, P' in the upper, and then to force the gluing along all sides, including S_1). Now $\bar{Q} \times SO(2)$ inherits a foliation from SL(2,R), (by upper triangular matrices) and this foliation is respected in the gluing procedure. One has to check that the foliation induced on the j-th boundary torus is conjugate to one by a line of slope α_j/π, and so one can fill in Reeb components to get a foliation of S^3.

The Godbillon-Vey class of this foliation is computed as before. Indeed, on the set Q-{vertices} $\times S^1$, modulo identifications, the form is simply $\theta \wedge \xi \wedge \eta$, and arguing as before then $\int_{S^3}$ G-V class of P $= \int_{SO(2)} \theta \int_Q \xi \wedge \eta = 4\pi$ (area of P) . Since one can get convex polygons in H of any prescribed area, the class takes on all real values as P varies.

Bott's examples. Take $M = C^2 - \{0\}$ and for each $A \in GL_2(C)$ define a holomorphic vector-field on M, $X_A = Az_1 \frac{\partial}{\partial z_1} + Az_2 \frac{\partial}{\partial z_2}$; here $Az_i = \sum_1^2 a_{ij}z_j$. X_A does not vanish on M: hence the subbundle $E_A \subseteq TM'$ which it spans is a 1-dimensional (over C) integrable subbundle; since M has complex dimension 2, E_A also has codimension 1, and in fact a trivial normal bundle, spanned for example by $\overline{Az}_2 \frac{\partial}{\partial z_1} - \overline{Az}_1 \frac{\partial}{\partial z_2}$.* The leaves of the foliation are simply the trajectories of X_A.

The holomorphic 1-form $w_A = Az_2\, dz_1 - Az_1\, dz_2$ is nowhere vanishing and annihilates X_A. Now $dw_A = -(\text{trace } A)\, dz_1 \wedge dz_2$ and if we put*

$$\theta_A = \frac{\text{trace } A}{\|Az\|^2} (\overline{Az}_1\, dz_1 + \overline{Az}_2\, dz_2) \text{ where } \|Az\|^2 = |Az_1|^2 + |Az_2|^2$$

we see at once that $dw_A = \theta_A \wedge w_A$. The Godbillon-Vey class is then represented by the 3-form $\theta_A \wedge d\theta_A$ which we can calculate as follows. First make a linear change of coordinates $\zeta_1 = Az_1$, $\zeta_2 = Az_2$ so that

$$\theta_A = \frac{\text{trace } A}{\|\zeta\|^2} (\bar{\zeta}_1\, dA^{-1}\, \zeta_1 + \bar{\zeta}_2\, dA^{-1}\, \zeta_2) .$$

Now if we put $f(\zeta) = \frac{\text{trace } A}{\|\zeta\|^2}$ and ϕ_A to be the remaining factor in θ_A,

$$\theta_A\, d\theta_A = f(\zeta)^2\, \phi_A\, d\phi_A = \frac{(\text{trace } A)^2}{\|\zeta\|^4} \frac{(\bar{\zeta}_2 d\zeta_1 d\zeta_2 d\bar{\zeta}_1 - \bar{\zeta}_1 d\zeta_1 d\zeta_2 d\bar{\zeta}_2)}{\det A}$$

where we have used $dA^{-1}\zeta_1 \wedge dA^{-1}\zeta_2 = (\det A^{-1})\, d\zeta_1 \wedge d\zeta_2$.

Now these forms certainly look different enough, for different A, but to show that they represent different cohomology classes as A varies, we must integrate $\theta_A\, d\theta_A$ over a 3-cycle in M. Set $S^3 = \{(\zeta_1, \zeta_2) \in C^2 : \|\zeta\|^2 = 1\}$. Then

---

*Since 0 is a removable singularity, it is easy to show that X_A cannot have a holomorphic complement, and similarly that even though θ_A is of type (1,0) it cannot be chosen to be holomorphic.

$$\int_{S^3} \theta_A \, d\theta_A = \frac{(\text{trace } A)^2}{\det A} \int_{S^3} (\bar{\zeta}_2 d\zeta_1 d\zeta_2 d\bar{\zeta}_1 - \bar{\zeta}_1 d\zeta_1 d\zeta_2 d\bar{\zeta}_2) .$$

It remains only to show that the integral on the right-hand side is not zero: this is a direct calculation which may safely be left to the reader.

Remarks. Note that the cohomology class of $\theta_A \, d\theta_A$ depends only on the conjugacy class of A, and furthermore is invariant under the substitution $A \to \lambda A$, $\lambda \in C^*$. This of course follows from the general property that the cohomology class is an invariant of the foliation. Under the substituion $A \to \lambda A$, although the vector-field is altered, the subbundle E_A is not. And the substitution $A \to T^{-1}AT$ corresponds to pulling back the foliation E_A by the (linear) diffeomorphism $T: C^2 - \{0\} \to C^2-\{0\}$. Since T^* is trivial on cohomology, the characteristic classes are the same.

II Higher exotic classes

1. The Godbillon-Vey Class in Codimension q

In order to motivate the general construction of the higher exotic classes for a foliation of codimension q, we begin by constructing a particular one which was announced by Godbillon and Vey in the same paper [10] as their codimension-one class. So consider an integrable subbundle $E \subseteq TM$ of codimension q, and assume that the normal-bundle $Q = TM/E$ is orientable. Then the projection $\pi: TM \to Q$ defines $\Lambda^q(\pi): \Lambda^q(TM) \to \Lambda^q(Q)$, and so for each choice of trivialization of $\Lambda^q(Q)$ we get a q-form on M; pick one, say Ω. The integrability of E implies that there is a 1-form $\psi \in A'(M)$ such that $d\Omega = \psi \wedge \Omega$ (see proof below) and again, although ψ is not unique, we have

Proposition 1. For any Ω and ψ as above, satisfying $d\Omega = \psi \wedge \Omega$, the form $\psi \wedge (d\psi)^q$ is closed, and its cohomology class $\{\psi \wedge (d\psi)^q\} \in H^{2q+1}(M)$ is a well-defined invariant of the integrable homotopy class of E.

Proof. We work in an open set $U \subseteq M$ over which there are q linearly independent 1-forms $\omega_1, \ldots, \omega_q$ annihilating E. Then we may take $\Omega|_U \quad \omega_1 \wedge \ldots \wedge \omega_q$. The integrability of E implies that there are 1-forms θ_{ij} such that

$$d\omega_i = \sum \theta_{ij} \wedge \omega_j \ .$$

Then

$$d\Omega = \sum (-1)^i \omega_1 \wedge \dots \wedge (d\omega_i) \wedge \dots \wedge \omega_q$$
$$= \sum (-1)^i \omega_1 \wedge \dots \quad (\sum \theta_{ij} \wedge \omega_j) \wedge \dots \wedge \omega_q$$
$$= \sum_i \theta_{ii} \wedge (\omega_1 \wedge \dots \wedge \omega_q) = (\text{trace } \Theta) \wedge \Omega$$

where we have put $\Theta = (\theta_{ij})$. For convenience, set $\theta = \text{trace } \Theta$ so $d\theta = \text{trace } (d\Theta)$.

Now recall from the proof of Theorem 1 that Θ is the connection matrix of some torsion-free connection on Q^*, and hence its curvature $k = d\Theta - \Theta \wedge \Theta$, evaluated on any invariant polynomial of degree > q is zero. We shall use this to show that $\theta \wedge (d\theta)^q$ is closed. First note that $d\theta = \text{trace } k$, because $\text{trace } (\Theta \wedge \Theta) = 0$. Now $d(\theta \wedge (d\theta)^q) = (d\theta)^{q+1} = (\text{trace } (k))^{q+1} = 0$, since $A \to (\text{trace } A)^{q+1}$ is an invariant polynomial of degree q+1.

Suppose ψ is some other one-form satisfying $d\Omega = \psi \wedge \Omega$: then $\psi = \theta + \alpha$ for some α such that $\alpha \wedge \Omega = 0$. From the identity $d\theta \wedge \Omega = 0$ one finds $\theta = \sum \eta_j \wedge \omega_j$ for some $\eta_j \in A^1(U)$ and the condition on α gives $\alpha = \sum f_j \omega_j$. Thus

$$(d\psi)^q = (d\theta)^q + \sum_{j=1}^{q} \binom{q}{j} (d\theta)^j (d\alpha)^{q-j}$$

and since $d\alpha$ is in the ideal generated by the ω_j, each $(d\theta)^j (d\alpha)^{q-j}$ is a multiple of Ω. Hence

$$\psi \wedge (d\psi)^q = \theta \wedge (d\psi)^q = \theta \wedge (d\theta)^q + \theta \wedge (\sum \dots)$$

and by a longer computation one can show the last term to be exact. Next if $\omega_1', \dots, \omega_q'$ is another basis of forms over U annihilating E, then $\Omega' = \omega_1' \wedge \dots \wedge \omega_q' = g\Omega$ where g is nowhere zero on M, and

$$d\Omega' = g\theta \wedge \Omega \pm dg \wedge \Omega = (\theta \pm \frac{dg}{g}) \wedge \Omega'$$

Setting $\phi = \theta \pm dg/g$, we have

$$\phi \wedge (d\phi)^q = (\theta \pm \frac{dg}{g}) \wedge (d\theta)^q = \theta \wedge (d\theta)^q \pm (\frac{dg}{g}) \wedge (d\theta)^q$$

and the second term is exact. The rest of the proof proceeds as in Theorem 2, Chapter 1.

We shall see later on that just as in the codimension one case this class is nonzero for various examples, and can even take on a continuum of values.

2. Real Secondary Classes. Let $V \to M$ be an arbitrary smooth vector bundle of rank q, equipped with two connections D^0 and D^1; let k^0, k^1 be their curvature tensors, and consider any invariant polynomial $\phi \in I(\underline{gl}_q)$. According to the Chern-Weil theory, the difference $\phi(k^0)-\phi(k^1)$ is an exact form. We shall now exhibit an explicit (2q-1) form ψ such that $d\psi = \phi(k^0)-\phi(k^1)$; this form will be useful for constructing secondary classes precisely because it is fashioned out of the connections D^0, D^1 and ϕ in a canonical manner.

To this end, pull back the bundle $V \to M$ over $M \times [0,1]$ by the obvious projection $\pi: M \times [0,1] \to M$. On $\pi^*(V)$, the operator $D = tD^1 + (1-t)D^0$ is a connection, which restricts at the two ends to D^0, D^1; let k be the curvature tensor of D. For any $\phi \in I(\underline{gl}_q)$ we can write $\phi(k)$ uniquely as $\phi(k) = \rho + \sigma\, dt$, where ρ does not involve dt. The operation* which drops the ρ and integrates $\sigma\, dt$ over [0,1] is well defined, and generally denoted by $\pi_*: A^p(M\times[0,1]) \to A^{p-1}(M)$. Then we have the following

*This is the usual "integration over the fibre"; for a rigorous treatment, consult any good book on differential geometry.

elementary

<u>Proposition 2</u>. $d_M(\pi_*(\phi(k))) = \phi(k^1) - \phi(k^0)$.

<u>Proof</u>. As remarked above, any form α on $M \times [0,1]$ can be written as $\alpha = \beta + \gamma\, dt$, where β does not involve dt, so that the restriction of α to any level $M \times \{t_0\}$ is just the restriction of β. We claim that

$$\pi_*(d\alpha) - d_M\pi_*(\alpha) = \alpha|_{M\times\{1\}} - \alpha|_{M\times\{0\}}$$

Let us compute the left-hand side

$$d_M\pi_*(\alpha) = d_M \int_0^1 \gamma\, dt = \int_0^1 (d_M\gamma)\, dt ,$$

$$d\alpha = d\beta + d(\gamma\, dt) = d_M\beta + \frac{\partial\beta}{\partial t} dt + (d_M\gamma)\, dt ,$$

so

$$\pi_*(d\alpha) = \int_0^1 \frac{\partial\beta}{\partial t}\, dt + \int_0^1 (d_M\gamma)\, dt .$$

Thus by the fundamental theorem of calculus (in one variable!) the identity above is verified.

In the case at hand, we know that $\phi(k)$ is a closed form on $M \times [0,1]$. So the identity becomes

$$d_M\pi_*(\phi(k)) = \phi(k^1) - \phi(k^0)$$

since $\phi(k)$ restricts to $\phi(k^0)$, $\phi(k^1)$ at the ends. □

To proceed with the construction of secondary classes, we need to record two further simple facts.

(1) The ring of invariant polynomials $I(\underline{gl}_q)$ is the polynomial

ring in q variables $R[c_1,\ldots,c_q]$, where the polynomials c_j, evaluated on a matrix $A \in \underline{gl}_q$ are given by $\det(t + A) = \sum_0^q c_i(A)t^{q-i}$. See Bott [2].

(2) If k is the curvature of a connection on a real vector bundle $V \to M$, then $c_{2i+1}(k)$ is exact. This follows from the fact that if we put a metric $\langle\ ,\ \rangle$ on V, we can define a Riemannian connection ∇ and the curvature k of ∇ is then skew-symmetric (see Appendix 1). Hence at any point x, the odd symmetric functions of the eigenvalues of the matrices $k_x(X,Y)$ are zero.

Now suppose that $E \subseteq TM$ is an integrable subbundle of codimension q, $Q^* \subseteq TM^*$ its annihilator. Let D^0 be a Riemannian connection on Q^* and let D^1 be a torsion-free connection on Q^*. Then we have three maps

$$\lambda_0, \lambda_1, : I(\underline{gl}_q) \to A^*(M)$$

given by

$$\lambda_i(\phi) = \phi(k^i), \qquad i = 0,1$$

$$\lambda(\phi) = \pi_*\phi(k)$$

where k^i = curvature of D^i, k is the curvature of $tD^1 + (1-k)D^0$ as above. Proposition 2 and Remark 2 following it show that

(i) $d\lambda(\phi) = \lambda_1(\phi) - \lambda_0(\phi)$

(ii) $\lambda_0(\phi) = 0$ if ϕ is a polynomial in the odd generators.

All this can be succinctly summarized by constructing a differential graded algebra WO_q and a map of DGA's $\tilde{\lambda}_E: WO_q \to A^*(M)$, as follows.*

As an algebra, $WO_q = \Lambda(u_1,u_3,\ldots,u_{2\ell+1}) \otimes R[c_1,\ldots,c_q]/$ (ideal of deg > q) where $\ell = [q/2]$. Here the degree of u_{2j-1} is $4j-3$ and the degree of c_j is $2j$. The differential is given by $du_{2i-1} = c_{2i-1}$, $dc_j = 0$. Note by our indexing conventions that only the "odd c_i" -- i.e., c_{2j-1} -- are exact. Define $\tilde{\lambda}_E: WO_q \to A^*(M)$ by requiring that

$$\tilde{\lambda}_E(u_i) = \lambda(c_i) = \pi_* c_i(k) , \qquad \tilde{\lambda}_E(c_i) = \lambda_1(c_i) = c_i(k^1)$$

and note that $\tilde{\lambda}_E$ does indeed vanish on the ideal defining WO_q. Moreover, $d\tilde{\lambda}_E(u_i) = d\pi_* c_i(k) = \lambda_1(c_i) - \lambda_0(c_i)$: as i is odd, $\lambda_0(c_i) = 0$ so $d\tilde{\lambda}_E(u_i) = \lambda_1(c_i) = \tilde{\lambda}_E(du_i)$: similarly since $c_i(k^1)$ is closed, $d\tilde{\lambda}_E(c_i) = \tilde{\lambda}_E(dc_i) = 0$. Thus $\tilde{\lambda}_E$ is a map of DGA's and so induces a map

$$\lambda_E: H^*(WO_q) \to H^*(M)$$

of cohomology algebras. The main result concerning λ_E is

<u>Theorem (Bott [2])</u>. λ_E is "functorial": that is, if $f: N \to M$ is a smooth map <u>transverse</u> to the foliation E then $f^* \circ \lambda_E = \lambda_{f^*(E)}$. In particular, λ_E is independent of the choice of the Riemannian and torsion-free connections (D^0,D^1) in the sense that if (∇^0,∇^1) is another pair, ∇^0 Riemannian with respect to some metric, ∇^1 torsion-free, then (∇^0,∇^1) defines the same map on cohomology. Also, λ_E depends only on the integrable-homotopy class of E.

---
*Similar algebras, called Weil algebras, enter naturally in one treatment of the usual Chern-Weil theory: see Cartan [5] and the Appendix.

Finally, $\{\lambda_E(c_1)\}$ = Pontryagin classes of Q^* and $\lambda_E(u_1 c_1^q)$ = Godbillon-Vey class of E.

Proof. We begin by "factoring" f as a composition $f\colon N \xrightarrow{i_f} M \times M \xrightarrow{\pi_M} M$ where $i_f(x) = (x, f(x))$ and π_M is projection on the second factor. Now π_M is clearly transverse to the subbundle $E \subseteq TM$, and so gives a codimension q foliation on $N \times M$. The condition of transversality on f then guarantees that this foliation on $N \times M$ is transverse to the imbedding i_f. Thus if we establish functoriality for maps (i) $g\colon X \to M$ which are transverse imbeddings, (ii) $h\colon Y \times M \to M$ which are projections on the second factor -- $h(y,m) = m$ then we will have functoriality for general transverse maps f. Thus it is simply a matter of checking that everything pulls back properly under maps like g, h above.

For any bundle $V \to M$, a connection on V gives a connection on the restriction $g^*(V) = V|_X$; moreover if the connection on V is Riemannian for some metric ρ, then the restricted connection is Riemannian for $g^*(\rho)$. The transversality of g guarantees that if $V = (TM/E)^*$ then $V|_X$ is the annihilator bundle of $E|_X$ and that a torsion-free connection restricts to a torsion-free one. So consider the case of a projection $h\colon Y \times M \to M$. For any bundle $V \to M$, $h^*(V) \simeq Y \times V$, so $\Gamma_{Y\times M}(h^*(V)) \simeq \Gamma_M(V) \otimes C^\infty(Y)$. Hence a connection $D\colon \Gamma_M(V) \to \Gamma_M(T\,M \otimes V)$ gives

$$\begin{array}{ccc} \Gamma_{Y\times M}(h^*(V)) \simeq C^\infty(Y) \otimes \Gamma_M(V) & \xrightarrow{d_Y \otimes D} & \Gamma_Y(T^*Y) \otimes \Gamma_M(T^*M \times V) \\ & \searrow^{D^1} & \wr\wr \\ & & \Gamma_{Y\times M}(T^*(Y\times M) \otimes h^*(V)) \end{array}$$

and D^1 is a connection. If $V = (TM/E)^*$ and D is torsion-free, then so is D^1, directly from the definition. Similarly if D is Riemannian for a metric ρ then D^1 is for $h^*(\rho)$.

Using the functoriality above, we shall prove the independence of λ_E of the pair (D^0,∇^0) used to construct $\tilde{\lambda}_E$. We state the principle which is then used repeatedly to get from a pair (D^0,∇^0) to a pair (D^1,∇^1). Consider the transverse map $\pi\colon M \times [0,1] \to M$, and suppose that on $\pi^*(Q^*)$ we have a family of torsion-free connections D_t, a family of metrics ρ_t and a family of connections ∇_t Riemannian with respect to ρ_t. Then the pair (D_t,∇_t) gives a map $\lambda_{\pi^*(E)}\colon H^*(WO_q) \to H^*(M\times[0,1])$ and if λ_0,λ_1 are the maps given by the pairs (D_0,∇_0) and (D_1,∇_1) on Q^*, then clearly $\lambda_0 = i_0^*\lambda_{\pi^*(E)}$, $\lambda_1 = i_1^*\lambda_{\pi^*(E)}$. By the transverse functoriality, $\lambda_{\pi^*E} = \pi^*\lambda_E$, and since $i_0^*\pi^* = i_1^*\pi^*$, $\lambda_0 = \lambda_1$.

This principle is used to pass from a pair (D^0,∇^0) to (D^1,∇^1) as follows. Let ρ, σ be metrics for which ∇^0,∇^1 are Riemannian. On $\pi^*(Q^*)$, $(1-t)\rho + t\sigma$ is a metric, and we can find <u>some</u> family ∇_t of Riemannian connections for it. We now apply the principle above three times. First we can pass from (D^0,∇^0) to (D^0,∇_0) because D^0 gives a torsion-free connection on $\pi^*(Q^*)$ and $t\nabla^0+(1-t)\nabla_0$ gives a Riemannian one for $\pi^*(\rho)$. We now pass from (D^0,∇_0) to (D^1,∇_1) by $D_t = (1-t)D^0 + tD^1$, ∇_t, and the metric $(1-t)\rho + t\sigma$; and finally from (D^1,∇_1) to (D^1,∇^1) by D^1, $(1-t)\nabla_1 + t\nabla^1$ and $\pi^*(\sigma)$. This three-stage argument is easier than constructing a 1-parameter family of Riemannian connections with prescribed metric, and prescribed end values.

The invariance of λ_E under integrable homotopy is very similar to the argument just given. The rest of the assertions in the theorem are trivial. □

Remarks. The cohomology algebra $H^*(WO_q)$ has been completely determined by J. Vey. See the exposé of Godbillon [9], as well as Chapter III, Section 2, for a brief summary. Salient features are:

(i) $H^n(WO_q) = 0$ for $n > q^2 + 2q$.

(ii) For $n \leq 2q$, $H^*(WO_q)$ = ring generated by universal Pontryagin classes.

(iii) There are many secondary classes in the range $2q+1 \leq n \leq q^2+2q$; for example, in $H^{2q+1}(WO_q)$ the classes u_1c^α where $c = \prod c_i^{\alpha_i}$ and $\sum \alpha_i = q$, are linearly independent.

(iv) The multiplicative structure is trivial in high degrees.

Corresponding to additional structures on the normal bundle, e.g. orientability, triviality, ..., one can construct associated truncated Weil algebras. In the next section we do this for holomorphic foliations, where the normal bundle has a complex structure.

Another important case is when the annihilator bundle Q^* is trivial, or more generaly has a flat Riemannian connection (i.e. the principal bundle of Q^* can be reduced to a finite covering). Then all the Riemannian forms c_i are zero, and one gets

$$\tilde{\mu}_E: W_q \to A^*(M)$$

where W_q is the DGA described as follows:

$$W_q = \Lambda^*(u_1, u_2, \ldots, u_q) \otimes R[c_1, \ldots, c_q]/(\text{ideal deg} > q)$$

with $\deg u_j = 2j-1$, $du_j = c_j$, $dc_j = 0$. In this case there are "more" exotic classes, compensating so to speak for the lack of any primary characteristic classes. This occurs in particular if we pull Q^* over its principal frame bundle $P(Q^*) \to M$; and then we can recover the secondary cohomology classes of Q^* from the diagram

$$\begin{array}{ccc} H^*(WO_q) & \xrightarrow{\tilde{\lambda}_E} & H^*(M) \\ \downarrow & & \downarrow \\ H^*(W_q) & \xrightarrow{\tilde{\mu}_E} & H^*(P(Q^*)) \end{array}$$

where the vertical injections are given by $SO(q)$-basic elements.

3. The Secondary Complex Classes

Suppose M is a complex analytic manifold, and E is an integrable subbundle of the holomorphic tangent-bundle TM' of complex codimension q. Theorem 1' asserts that the Chern-ring of TM'/E is zero in degrees greater than 2q, so one can construct secondary classes in this setting as well. The general pattern is the same as in the real case, so we shall be brief.

First let $V \to M$ be any complex vector bundle. Then V admits a Hermitian metric, and with respect to it, a "Hermitian connection" which satisfies an identity analogous to that of a Riemannian connection in the real case. The curvature matrix of this connection is skew-Hermitian, so the Chern forms are in fact real (one multiplies the curvature by $i/2\pi$). This leads to a relation among the Chern forms $c_i(\nabla)$ is an arbitrary connection, namely that $c_i(\nabla) - \bar{c}_i(\nabla)$ is exact. In precise analogy with proposition 2, one can then find canonical forms $u_i(\nabla,<\ ,\ >)$ constructed from the given connection and the Hermitian one such that $du_i(\nabla,<\ ,\ >) = c_i(\nabla) - \bar{c}_i(\nabla)$.

Thus if we apply this to $Q' = TM'/E$ and two connections D^0, D^1 on $Q;$, D^0 being Hermitian and D^1 being torsion-free, we get three maps

$$\lambda_0, \lambda_1, \lambda\colon I(\underline{gl}_q(C)) \to A_C^*(M)$$

given as before by

$$\lambda_i(\phi) = \phi(k^i) \ , \quad k^i = \text{curvature of } D^i$$
$$\lambda(\phi) = \phi(u_i) \ , \quad \text{where } u_i \text{ is as above.}$$

The appropriate DGA for the secondary classes is then constructed as follows:

$$WU_q = A_C(u_1, u_2, \ldots, u_q) \otimes \frac{C[c_1, \ldots, c_q]}{(\text{ideal deg} > 2q)} \otimes \frac{C[\bar{c}_1, \ldots, \bar{c}_q]}{(\text{ideal} > 2q)}$$

with $du_i = c_i - \bar{c}_i$, $dc_i = 0$ (and hence $d\bar{c}_i = 0$ as well). The three maps constructed above combine to give a map of DGA's

$$\lambda_E\colon WU_q \to A_C^*(M)$$

by

$$\lambda_E(u_i) = \lambda(u_i)$$

$$\lambda_E(c_i) = \lambda_0(c_i) \ .$$

Theorem. The induced map $\lambda_E^*\colon H^*(WU_q) \to H_{DR}^*(M:C)$ is independent of the choice of the pair (D^0, D^1) in the sense that if (∇^0, ∇^1) is another pair, ∇^0 Hermitian for some metric, ∇^1 torsion-free, then (∇^0, ∇^1) defines the same map on cohomology; and λ_E is an invariant of the transverse homotopy class of E.

If $f: N \to M$ is transverse to the foliation, $f^* \circ \lambda_E = \lambda_{f^*(E)}$. Finally $\lambda_E(c_i)$ = Chern classes of Q'.

The proof is similar to the proof of the theorem in the previous section, and the transverse homotopy invariance of λ_E follows as before. The cohomology algebra $H^*(WU_q)$ has also been computed by J. Vey; see Godbillon [9] again. As in the real case, there are many exotic classes in degrees between 2q+1 and $2q^2+q$, and there is a holomorphic Godbillon-Vey class $u_1 c_1^q$.

In the next section we shall see that the classes $u_1 c^\alpha$ in degrees 2q+1 can take on all complex values.

4. Examples

We now give some computations of the higher exotic classes constructed above, which show that they are not zero. All the examples are related to semisimple Lie groups and their homogeneous spaces and there ought to be a general result in these terms. In fact Theorem 2 of [3] is such a result, but unfortunately it is not quite correct (as Professor Bott pointed out to me).

The foliations of Roussarie on $SL(2,R)/\Gamma$ can be generalized in two ways, because the families $SL(n,R)$ and $SO(n,1)$ coincide for $n = 2$. Each generalization gives the nontriviality of a higher exotic class, and these are the first two examples below.

(a) For each integer $n \geq 2$, there exist compact, oriented manifolds S_n of constant negative curvature* (for $n = 2$ these

*See for example Wolf [22] for a general discussion.

are precisely the compact Riemann surfaces of genus ≥ 2). Let M_n = unit tangent sphere-bundle of S_n. On M_n there is a natural (Anosov) foliation of codimension (n-1); its leaves are comprised of all tangent vectors whose geodesics are parallel, in the sense of hyperbolic geometry. For this foliation we shall show that $u_1c_1^{n-1} \in H^{2n-1}(M)$ is not zero.

Just as with Riemann surfaces, these S_n have a common universal covering, the so-called hyperbolic space* $\mathcal{H}_n$ which can be described as a component of a cone in R^{n+1}: $\mathcal{H}_n = \{(x_0,\ldots,x_n) \in R^{n+1}: x_0 > 0, -x_0^2 + \sum_1^n x_i^2 = -1\}$. Let SO(n,1) be the subgroup of orientation preserving linear maps which preserve the quadratic form $-x_0^2 + \sum_1^n x_i^2$. SO(n,1) acts transitively on $\mathcal{H}_n$ and the isotopy subgroup of (1,0,...,0) is SO(n). Thus SO(n,1)/SO(n) $= \mathcal{H}_n$ and every discrete, torsion-free subgroup $\Gamma \subseteq$ SO(n,1) with compact quotient gives a compact manifold $\Gamma \backslash$ SO(n,1)/SO(n) $= S_n$ as above. (The existence of such Γ was known in this case, but there is also a general result of Borel [1] to this effect.) The tangent sphere-bundle of $\mathcal{H}_n$ can then be regarded as SO(n,1)/SO(n-1) and so the M_n above are all of the form $\Gamma \backslash$ SO(n,1)/SO(n-1). The foliation described above is obviously left invariant under isometrics, and so we will work on SO(n,1)/SO(n-1).

To describe the foliation in this setting, we note that it comes from looking at the action of SO(n,1) on the "boundary" of $\mathcal{H}_n$; when n = 2, we get the foliation by left cosets of the upper triangular matrices in SL(2,R), and this is precisely the

*See for example Wolf [22] for a general discussion.

foliation coming from the action of $SL(2,R)$ on $S^1 = \partial$ (unit disk). The point is that every parallel family of geodesics converges to a given point on the boundary, and every point on the boundary represents such a parallel family.

We now set about the computation. Let $Q_n = \{[z_0,\ldots,z_n] \in P_n(C) : -z_0^2 + \sum_1^n z_i^2 = 0\}$ and $Q_n(R) = Q_n \cap P_n(R)$ be the real points. It is trivial to see that $Q_n(R) \simeq S^{n-1}$; (on $Q_n(R)$, x_0 cannot be zero). This is the boundary of H_n referred to earlier. $SO(n,1)$ acts on $Q_n(R)$ transitively: let P be the isotropy group of the point $[1,0,\ldots,0]$. It is easy to see that $SO(n-1) \subseteq P$ and hence that we have a fibre-bundle $P/SO(n-1) \to SO(n,1)/SO(n-1) \xrightarrow{\pi} SO(n,1)/P$ which foliates the total space by the fibres. The annihilator bundle of this foliation is of course $\pi^*(T^*Q_n(R))$. The cotangent bundle $T^*Q_n(R) \to Q_n(R)$ has a natural homogeneous connection on it, and this gives a torsion-free connection on the pull-back.

In the formalism of Cartan [5], this homogeneous connection gives a map of the Weil algebra $W(O(n-1))$ into the <u>invariant</u> forms on the space: $\psi: W(O(n-1)) \to \Lambda^*(\underline{so}(n,1),\underline{so}(n-1))$ * (see Appendix I). As the connection is torsion-free, ψ factors through the truncated Weil algebra to give $\psi: WO_{n-1} \to \Lambda^*(\underline{so}(n,1),\underline{so}(n-1))$ and thus a map on cohomology $\psi^*: H^*(WO_{n-1}) \to H^*(\underline{so}(n,1),\underline{so}(n-1))$. We shall first show that $\psi^*(u_1c_1^{n-1}) \neq 0$, and then relate the Lie algebra cohomology to $H^*(M_n)$.

* Given a Lie algebra $\underline{g}$ and subalgebra $\underline{h}$ we write $\Lambda^*(\underline{g},\underline{h})$ for the complex $\Lambda^*((\underline{g}/\underline{h})^*)$.

To compute $\psi^*(u_1c_1^{n-1})$ we pass to the complex case because here the Lie algebra cohomology is the de Rham cohomology (see Appendix II) so that we can throw the brunt of the computation on known topological facts. Thus we consider the fibre-bundle $P(C)/SO(n-1:C) \to SO(n+1:C)/SO(n-1:C) \xrightarrow{\pi} Q_n(C)$ (recall that $SO(n,1:C) = SO(n+1:C)$) which foliates the total space here, and that the real-bundle is sitting inside the complex one in the obvious way. The homogeneous connection on $T^*Q_n(C)$ is just the complex extension of the one in $T^*Q_n(R)$ and the Cartan map is then $\phi\colon WU_n \to \Lambda^*(\underline{so}(n+1;C), \underline{so}(n-1;C))$. In fact the diagram

$$\begin{array}{ccc} WO_n & \xrightarrow{\psi} & \Lambda^*(\underline{so}(n,1),\underline{so}(n-1)) \\ \downarrow & & \downarrow \otimes C \\ WU_n & \xrightarrow{\phi} & \Lambda^*(\underline{so}(n+1;C),\underline{so}(n-1;C)) \end{array}$$

commutes, and as the right vertical is just $\otimes$ C on cohomology as well, $\psi(u_1c_1^{n-1}) \neq 0$ if and only if $\phi(u_1c_1^{n-1}) \neq 0$. Now as remarked above, $H^*(\underline{so}(n+1;C,\underline{so}(n-1;C)) \simeq H^*_{DR}(SO(n+1;C)/SO(n-1;C)$ so it suffices to show that $\phi(u_1c_1^{n-1}) \neq 0$ in the de Rham cohomology on the right-hand side. For this purpose we change our ground once more. Let $V_n \subseteq C^{n-1} - \{0\}$ be the C^*-cone over Q_n: $C^* \to V_n \xrightarrow{\pi} Q_n$. Then V_n is foliated by the C^*-orbits and we can work with this foliation without destroying any information, because of the following commutative diagram

$$\begin{array}{ccccc} P(C)/SO(n-1;C) & \to & SO(n+1;C)/SO(n-1;C) & \xrightarrow{\pi} & Q_n \\ \downarrow & & \downarrow & & \| \\ C^* & \to & V_n & \xrightarrow{\pi} & Q_n \end{array}$$

The vertical maps are bundle maps, and as their fibres are contractible, they are homotopy-equivalences. Moreover, the foliation on the top total space is just the pull-back of the foliation on V_n as one can see in several ways -- for example because the annihilator-bundle in both cases is $\pi^*(T^*Q_n)$.

Let us interpret $\phi(u_1c_1^{n-1}) \in H^{2n-1}(V_n)$ geometrically. Recall that our basic onnection is the pull-back of a connection on T^*Q_n, so even at the level of differential forms, $\phi(c_1) = \pi^*(c_1(T^*Q_n))$ and $\phi(u_1)$ is a transgressed form which generates $H^1(C^*)$ in every fibre. The argument is now completed as follows: $c_1 \in H^2(Q_n)$ is a generator, and $c_1^{n-1} \in H^{2n-2}(Q_n)$ is a nonzero multiple of the volume form.* From the Gysin sequence, $H^{2n-1}(V_n) \cong H^{2n-2}(Q_n)$ and a generator of H^{2n-1} is represented by a form $\omega_1 \wedge \pi^*(\omega_2)$ where ω_1 restricts a generator of H^1 in every fibre, ω_2 represents a volume form on Q_n. Thus $\phi(u_1)\phi(c_1)^{n-1}$ is a nonzero multiple of $\omega_1 \wedge \pi^*(\omega_2)$, so nonzero.

We still have to show $u_1c_1^{n-1} \in H^{2n-1}(M_n)$ is nonzero. Now semi-simple Lie algebras satisfy Poincaré duality in Lie algebra cohomology (Koszul [16]) so $\psi(u_1c_1^{n-1}) \in H^*(so(n,1),so(n,1)$ is a multiple of a "volume form". This descends to a volume form on any $\Gamma \backslash SO(n,1)/SO(n-1)$, and so by the compactness of M_n, we are done.

(b) The second generalization of Roussarie's examples is the left-invariant foliation of SL(n+1,R) given by cosets of the

*Note that Q_n is a complex hypersurface in $P_n(C)$, so has real dimension 2n-2.

parabolic subgroup $P = \{(a_{ij}) \in SL(n+1,R) \mid a_{00}>0,\ a_{k0}=0,$ $k = 1,2,\dots,n\}$; geometrically this foliation "comes from" the transitive action of $SL(n+1,R)$ on S^n = space of rays in R^{n+1}, since P is the isotropy subgroup of the point $(1,0,\dots,0) \in S^n$. Since the foliation is left-invariant, the computations of secondary classes can be done entirely with left-invariant forms on $G = SL(n+1,R)$ and then transferred to any quotient space $\Gamma \backslash G$, where $\Gamma \subseteq G$ is a discrete subgroup. Here we shall work with the left-invariant complex $\Lambda^*(\underline{sl}^*_{n+1})$ and we shall show that in the top dimension, $(u_1\, u_3 \dots u_{2n-1})c^\alpha \in H^{n^2+2n}(\underline{sl}_{n+1})$ is nonzero, c^α being a Chern monomial $c_1^{\alpha_1} \dots c_j^{\alpha_j}$ with $\sum i_j\alpha_j = n$. Thus if $\Gamma \subseteq SL(n+1,R)$ is co-compact (such subgroups exist by Borel [1]) then $(u_1 \dots u_{2n-1})c^\alpha \in H^{n^2+2n}(\Gamma\backslash SL(n+1:R):R)$ will be a nonzero multiple of the volume form. This example has been claimed (independently) by various mathematicians, but again without any published proofs.

Now $\underline{sl}^*_{n+1} = \{\omega_{ij} \mid 0 \le i,j \le n,\ \sum_0^n \omega_{jj} = 0\}$ and because $\underline{sl}_{n+1} \subseteq \underline{gl}_{n+1}$, $d\omega_{ij} = \sum_{k=0}^{n} \omega_{ik} \wedge \omega_{kj}$. The forms generating the dual of the normal bundle are those which annihilate the subalgebra $\underline{p}$ of P: a basis for these forms is clearly $\{\omega_{k_0} \mid k = 1,2,\dots,n\}$. These facts allow use to compute the Bott connection Θ and its curvature Ω, and then the Chern polynomials c^α and the transgressive forms u_i. Actually, we shall compute the trace powers $\psi_j = \text{trace}\ (\Omega^j)$ and classes $h_j \in \Lambda^{2j-1}(\underline{sl}^*_{n+1})$ associated to Θ such that $dh_j = \psi_j$: and we shall show that $(h_1\, h_2 \dots h_n)\psi^\alpha$ is nonzero in the top dimension, where as above $\psi^\alpha = \psi_1^{\alpha_1} \dots \psi_{i_k}^{\alpha_k}$, $|\alpha| = \sum i_j\alpha_j = n$. By Newton's formulae, suitably applied, this is equivalent to the nonvanishing of the classes $(u_1 \ldots u_{2n-1})c^{\alpha}$.

We begin by computing the Bott connection Θ.

$$d\omega_{j0} = \omega_{j0} \wedge \omega_{00} + \sum_1^n \omega_{jk} \wedge \omega_{k0} = \left(\sum_1^n \omega_{kk}\right) \wedge \omega_{j0} + \sum_{k=1}^n \omega_{jk} \wedge \omega_{k0}$$

So

$$\Theta = W + \sigma I_n , \quad W = (\omega_{ij}) , \qquad 1 \leq i,j \leq n$$

$$\sigma = \sum_1^n \omega_{kk} = \text{trace } (W)$$

Thus

$$\Omega = d\Theta - \Theta \wedge \Theta = (dW - W \wedge W) + (d\sigma) I_n$$

because $$W \wedge \sigma + \sigma \wedge W = 0 .$$

Let $V = dW - W \wedge W$, and $\tau = d\sigma$: note that $\tau = \text{trace } (dW)$ $= \text{trace } (V)$. The entries of V take a very simple form: the (i,j)th entry of dW is given by

$$d\omega_{ij} = \omega_{i0} \wedge \omega_{0j} + \sum_{k=1}^n \omega_{ik} \wedge \omega_{kj}$$

and the second summand is precisely the (i,j)th entry of $W \wedge W$. So

$$\Omega = V + \tau I_n , \quad V = (\omega_{i0} \wedge \omega_{0j}) , \quad \tau = \text{trace } V .$$

We can now compute the trace powers $\psi_r = \text{trace } (\Omega^r)$, $r \geq 1$. First of all, Ω is a matrix of 2-forms, and since the second summand is a scalar, we have

$$\Omega^r = \sum_{s=0}^r \binom{r}{s} V^s \tau^{r-s}$$

whence trace $(\Omega^r) = n\tau^r + \sum_{s=1}^{r} \binom{r}{s}$ trace $(V^s) \cdot \tau^{r-s}$.

Lemma. trace $(V^s) = (-1)^{s-1} \tau^s$.

Proof. This is a formal consequence of the fact that V is the tensor product of an $(n\times 1)$-matrix by an $(1\times n)$-matrix. Explicitly, set $\alpha_i = \omega_{i0}$, $\beta_j = \omega_{0j}$ so that $V = (\alpha_i \wedge \beta_j)$. Then the $(i,j)^{th}$ entry of V^s is

$$\sum \alpha_i \wedge \beta_{i_1} \wedge \alpha_{i_1} \dots \wedge \beta_{i_{s-1}} \wedge \alpha_{i_{s-1}} \wedge \beta_j$$

summed (freely) over all (s-1)-tuples $(i_1,\dots,i_{s-1})$. Thus

$$\text{trace } (V^s) = \sum \alpha_i \wedge \beta_{i_1} \wedge \dots \wedge \alpha_{i_{s-1}} \wedge \beta_i$$

summed (freely) over all s-tuples $(i,i_1,\dots,i_{s-1})$. Now bringing the β_i in next to the α_i , and switching each $\beta_{i_j} \wedge \alpha_{i_j}$ around, we get the result. □

It follows that trace $(\Omega^r) = n\tau^r + \left(\sum_{s=1}^{r} \binom{r}{s} (-1)^{s-1}\right)\tau^r$, and since the binomial coefficient is 1, ψ_r =trace (Ω^r) $= (n+1)\tau^r$. The point to keep in mind is that each ψ_r is a constant multiple of $\psi_1 = c_1$: indeed, the constant is $(n+1)^{1-r}$.

Next we compute the forms $h_j \in \Lambda^{2j-1}(\underline{sl}^*_{n+1})$ associated to Θ which satisfy $dh_j = \psi_j$. The fact that $h_1 = (n+1)\sigma$ is clear, since $(n+1)\sigma$ is the unique one-form in $\Lambda^1(\underline{sl}^*_{n+1})$ whose exterior derivative is $(n+1)\tau = \psi_1$; since the normal bundle on $SL_{n+1}(R)$ is trivial, the higher h_j's are obtained by a formal "integration over the fibre" (see Chapter III, Section 2). This procedure is carried out as follows: set $\Theta_t = t\Theta$, and Ω_t to be the associated curvature form,

$$\Omega_t = d\Theta_t - \Theta_t \wedge \Theta_t = t\Omega + (t-t^2)\Theta \wedge \Theta + \Theta \wedge dt \ .$$

Then $h_j = \pi_*(\text{trace } (\Omega_t^j))$ where π_* means "pick the dt-terms, and integrate from 0 to 1". The first computational difficulty here is that Ω and Θ do <u>not</u> commute (for $n \geq 2$), so one cannot use the commutative binomial theorem on $t\Omega$ and $A = (t-t^2)\Theta^2 + \Theta\, dt$ to calculate the powers Ω_t^j. However, we are not really interested in explicit expressions for the h_j themselves: we only want to compute certain products $h_{i_1} \dots h_{i_k} \psi^\alpha$, $|\alpha| = n$. Now, up to nonzero constants, $\psi^\alpha = \psi_1^n$ and one checks $\psi_1^n = \pm (\omega_{10} \wedge \dots \wedge \omega_{n0}) \wedge (\omega_{01} \wedge \dots \wedge \omega_{0n})$: moreover, it is tautologically true that

$$\Omega_t^r = \sum \binom{r}{s} t^r \Omega^r A^{r-s} \quad \text{mod (ideal gen by } [\Omega,\Theta])$$

and each entry of $[\Omega,\Theta]$ is in the ideal generated by $\{\omega_{i0}, \omega_{0j} | 1 \leq i,j \leq n\}$. Thus

$$\text{trace } (\Omega_t^r) = \sum \binom{r}{s} t^r \text{ trace } (\Omega^s A^{r-s}) + \eta_r$$

where η_r is in the ideal generated by $\{\omega_{i0}, \omega_{0j} | 1 \leq i,j \leq n\}$. Hence in mputing $h_{i_1} \dots h_{i_k} \psi^\alpha$ we may as well take $h_r \sim \pi_*\left[\sum \binom{r}{s} t^r \text{ trace } (\Omega^s A^{r-s})\right]$. A little honest work (remembering that we are only interested in the dt coefficient of trace (Ω_t^r)) then gives

$$h_r \sim \sum_{s=0}^{r} a_s \text{ trace } (\Omega^{r-s}\, \Theta^{2s-1}) \quad \text{with } a_s > 0 \text{ for all } s.$$

Finally we can compute $h_1 \dots h_{2n-1} \psi^\alpha$, $|\alpha| = n$. Again, in the above expression for h_r , all summands except the one for $s = r$ are in the ideal generated by $\{\omega_{i0}, \omega_{0j} | 1 \leq i,j \leq n\}$ so we can make the further reduction

$$h_r \sim a_r \text{ trace } (\Theta^{2r-1})$$

so that we get

$$\left(\prod_1^n h_r\right)\psi^\alpha = \left(\prod_1^n a_r\right)\left[\prod_{r=1}^n \text{trace } (\Theta^{2r-1})\right]\psi^\alpha$$

To compute the product of the traces on the right-hand side we avail ourselves of yet another reduction: recall that $h_1 = (n+1)\sigma$ and $\Theta = W + \sigma I_n$ so $\Theta^j = \sum_{s=0}^{j} \binom{j}{s}(-1)^{j-s} W^{j-s} \sigma^s$ and

$$\text{trace } (\Theta^j) = \text{trace } (W^j) + \sigma(\sum \ldots)$$

Thus finally

$$\left(\prod_1^n h_r\right)\psi^\alpha = a(n+1)\left[\prod_{r=1}^n \text{trace } (W^{2r-1})\right]\psi^\alpha$$

where $a = \prod_1^n a_r$ is a positive constant. Now the product of the traces of the W^{2r-1} falls out of general, theoretical facts W is the Maurer-Cartan matrix of $\underline{gl}_n \leq \underline{sl}_{n+1}$, and so the forms trace (W^{2r-1}) constitute a basis for the adjoint invariants $\Lambda^*(\underline{gl}_n^*)^{\underline{gl}_n}$: as this is an exterior algebra on these n generators, the product of the traces is (at least up to sign) the volume form: $\prod_{r=1}^n \text{trace } (W^{2r-1}) = \pm (\omega_{11} \wedge \ldots \omega_{1n} \wedge \omega_{21} \wedge \ldots \wedge \omega_{2n} \wedge \ldots \omega_{nn})$. Thus we see that

$$(\prod h_j)\psi_\alpha = c(\text{volume form on}$$

$c = \pm a(n+1) \prod (n+1)^{1-\alpha_i}$, and this fulfills our claim.

One can continue along the same lines to show that all the classes $h_1 c^\alpha$ are so(n)-basic and give nonzero multiples of the (transgressed) Euler class in $H^{2n+1}(\underline{sl}_{n+1}, \underline{so}(n+1))$.

(c) Let $M = C^{n+1} - \{0\}$. Each $A \in GL_{n+1}(C)$ defines a non-vanishing holomorphic vector-field $X_A = \sum_0^n (Az)_j \frac{\partial}{\partial z_j}$ where as usual $(Az)_j = \sum_{k=0}^n a_{jk} z_k$. This then gives a trivial subbundle $E_A \subseteq T^{1,0}M$ of complex codimension n, which is integrable because it is one-dimensional. In this case all the classes $u_1 c^\alpha \in H^{2n+1}(M:C)$ vary smoothly with A; we shall give the details for $u_1 c_1^n$ and discuss the other cases briefly. Except for a misprint in their formula, this is essentially Theorem 3 of Bott and Haefliger [3].

We begin with some remarks. The first point is that in contrast with the case n = 1, for general n the bundle $Q_A^* = (T^{1,0}M/E_A)^*$ is <u>not</u> trivial: indeed if there were n global linearly independent (1,0) forms $\xi_1, \ldots, \xi_n$ such that $\xi_i(X_A) = 0$ (even of class C^0), we would have n linearly independent vector fields of the same differentiability class on the sphere $S_A^{2n+1} = \{(z_0, \ldots, z_n) \in n \mid \sum |Az_j|^2 = 1\}$. But the Radon-Hurwitz number for S^{2n+1} is in general much smaller than n.

The second remark is that when A is theidentity matrix, this example essentially reduces to the complex analogue of example b. For X_I is the radial vector-field, and our foliation is then given by the fibres in the bundle $\underset{\sim}{C}^* \to M \overset{\pi}{\to} \underset{\sim}{P}_n$. (Here $\underset{\sim}{P}_n = \underset{\sim\sim}{CP}_n$). Now M is a homogeneous space of $SL_{n+1}(\underset{\sim}{C})$, so there is a bundle map $\rho: SL_{n+1}(\underset{\sim}{C}) \to M$, and the composite $\sigma = \pi \circ \rho$ then describes the fibre-bundle

$$P \longrightarrow SL_{n+L}(\underset{\sim}{C}) \xrightarrow{\sigma} \underset{\sim}{P}_n$$

which gives the foliation of $SL_{n+1}(\underset{\sim}{C})$ by cosets of P. Moreover, the map $\rho^*: H^*(M:\underset{\sim}{C}) \to H^*(SL_{n+L}(\underset{\sim}{C}): \underset{\sim}{C}) \simeq H^*(sl_{n+1}:\underset{\sim}{C})$ is injective, and so we could carry out this computatin in lie-algebra cohomology, using invariant forms. The reader is invited to check that the calculation of the classes u_1c^α given below then agrees with $\rho^*(u_1c^\alpha)$ as given in example b. Now even for general A, we do have a foliation on SL_{n+1}, which is however not homogeneous except at the "critical point" A = I: this point of view will be carried out elsewhere.

To return to the foliation on M, the last point to note is that even if $Q_A^* = (T^{1,0}M/E_A)^*$ is not trivial, its top exterior power $\Lambda^n_{\underset{\sim}{C}}(Q_A^*)$ is trivial: this follows from the fact that E_A and $T^{1,0}M$ are both trivial bundles, or from the fact that $c_1(Q_A^*) \in H^2(M) = 0$. This makes the computation of $u_1c_1^n$ much simpler than that of the other classes u_1c^α, and so we begin with a full discussion of the calculation of the Godbillon-Vey invariant, $u_1c_1^n$.

The projection $T^{1,0}M \to Q_A$ gives a map on the nth exterior powers, and choosing a trivialization of $\Lambda^n_{\underset{\sim}{C}}(Q_A)$ gives a holomorphic n-form Ω_A such that $d\Omega_A = \psi \wedge \Omega_A$: we already know $u_1c_1^n = \{\psi \wedge (d\psi)^n\}$ and that this is independent of all the choices involved.

For notational convenience, set $\zeta_j = (Az)_j = \sum a_{jk}z_k$, and note that $(\zeta_0,\ldots,\zeta_n)$ is a holomorphic coordinate system on M. The (1,0)-form $\xi = \sum_0^n \bar{\zeta}_j\, dz_j$ is nowhere vanishing and

$\xi(X_A) = \sum_0^n |\zeta_j|^2 \neq 0$. We claim that we can choose one ψ to be $f\xi$, where f is some function to be determined presently. Indeed, if $U \subseteq M$ is an open set over which Q_A^* is spanned by $\omega_1,\ldots,\omega_n$ then $\{\xi,\omega_1,\ldots,\omega_n\}$ is a basis for the space of (1,0) forms over U, and Ω_A and $\omega_1 \wedge \ldots \wedge \omega_n$ are proportional over U. Writing $\psi|_U = f_U\xi + \sum g_j^U\omega_j$ we see that $(\psi \wedge \Omega_A)|_U = f_U\xi \; (\Omega_A|_U)$, and since $\psi \wedge \Omega_A$, $\xi \wedge \Omega_A$ are global (n+1,0)-forms, the f_U must agree in the overlaps to give a global f such that $\psi \wedge \Omega_A = f\xi \wedge \Omega_A$.

Our task then is to determine f explicitly and to calculate $\psi \wedge (d\psi)^n$ in terms of f and ξ. The second part is easy.

$$(d\psi)^n = (df \wedge \xi + f\, d\xi)^n = f^n\,(d\xi)^n \bmod \xi$$

so

$$\psi \wedge (d\psi)^n = f^{n+1}(\xi \wedge (d\xi)^n) .$$

Now $d\xi = \sum_0^n d\bar{\zeta}_j \wedge dz_j$ and the forms $\{dz_0,\ldots,dz_n,d\bar{\zeta}_0,\ldots,d\bar{\zeta}_n\}$ are a basis for TM. We find then

$$\xi \wedge (d\xi)^n = \sum_0^n (-1)^{n+j}\,\bar{\zeta}_j(dz_0 \wedge \cdots \wedge dz_n) \wedge (d\bar{\zeta}_0 \wedge \ldots \wedge \widehat{d\bar{\zeta}_j} \wedge \ldots d\bar{\zeta}_n)$$

and written entirely in ζ-coordinates

$$\xi \wedge (d\xi)^n = \frac{1}{\det A}\sum (-1)^{n+j}\,\bar{\zeta}_j(d\zeta_0 \wedge \ldots \wedge d\zeta_n) \wedge (d\bar{\zeta}_0 \wedge \ldots \wedge \widehat{d\bar{\zeta}_j} \wedge \ldots \quad d\zeta_n)$$

which, apart from the factor of det A, is the "standard" volume form in the ζ-unit sphere.

To determine f explicitly we choose an open set $U \subseteq M$ over which Q_A^* is trivial: this allows us to compute $d\Omega_A$ explicitly, and comparing this form with $\xi \wedge \Omega_A$ we will get $f|_U$. In principle we would have to do this over an open cover of M: fortuitously, Q_A^* is trivial over the Zariski open set $U = \{(\zeta_0,\ldots,\zeta_n) \in M \mid \zeta_0 \neq 0\}$ and so $f|_U$ will determine f -- in fact, the formula for $f|_U$ will make sense over M, and this then gives f.

Consider the n (1,0)-forms given by

$$\omega_j = \zeta_j\, dz_0 - \zeta_0\, dz_j \,, \quad 1 \leq j \leq n.$$

These forms annihilate X_A and their product is

$$\omega_1 \wedge \ldots \wedge \omega_n = (\zeta_0)^{n-1}\left(\sum (-1)^k \zeta_k\, dz_0 \wedge \ldots \wedge \widehat{dz_k} \wedge \ldots \wedge dz_n\right)$$

Thus on the set $U = \{(\zeta_0,\ldots,\zeta_n) \mid \zeta_0 \neq 0\}$, $\omega_1,\ldots,\omega_n$ is a basis for $Q_A^*|_U$, and we can take $\Omega_A|_U$ to be any multiple of $\omega_1 \wedge \ldots \wedge \omega_n$. We choose to set

$$\Omega_A|_U = (\zeta_0)^{1-n}\, \omega_1 \wedge \ldots \wedge \omega_n = \sum (-1)^k \zeta_k\, dz_0 \wedge \ldots \widehat{dz_k} \ldots \wedge dz_n$$

Then computing its derivative, we find

$$\begin{aligned} d\Omega_A|_U &= \sum (-1)^k d\zeta_k \wedge dz_0 \wedge \ldots \widehat{dz_k} \ldots \wedge dz_n \\ &= \text{trace } (A)\ (dz_1 \wedge \ldots \wedge dz_n) \end{aligned}$$

because only the dz_k-component of $d\zeta_k$ contributes to the sum. Let us compare this with $(\xi \wedge \Omega_A)|_U$:

$$\begin{aligned} (\xi \wedge \Omega_A)|_U &= \left(\sum \bar{\zeta}_j\, dz_j\right) \wedge \left(\sum (-1)^k \zeta_k\, dz_0 \wedge \ldots \widehat{dz_k} \ldots \wedge dz_n\right) \\ &= \left(\sum |\zeta_j|^2\right) (dz_0 \wedge \ldots \wedge dz_n) \end{aligned}$$

and so we find $f(\zeta_0,\ldots,\zeta_n) = (\text{trace } A)/\|\zeta\|^2$

Thus finally we have the Godbillon-Vey form

$$\psi \wedge (d\psi)^n = \frac{(\text{trace } A)^{n+1}}{\|\zeta\|^{2n+2}} (\xi \wedge (d\xi)^n) =$$

$$= \frac{(\text{trace } A)^{n+1}}{\det A} \cdot \frac{1}{\|\zeta\|^{2n+2}} \sum_{j=0}^{n} (-1)^{n+j} \bar{\zeta}_j (d\zeta_0 \wedge \cdots \wedge d\zeta_n)$$

$$(d\bar{\zeta}_0 \wedge \ldots \wedge \widehat{d\bar{\zeta}_j} \ldots \wedge d\bar{\zeta}_n)$$

Integrating this over the ζ-unit sphere, the power of the norm drops out and we have

$$\int_S \psi \wedge (d\psi)^n = \frac{(\text{trace } A)^{n+1}}{\det A} c$$

where c is a "universal constant" depending on the volume of the sphere S^{2n+1} with respect to the standard volume form given above.

To calculate the other classes $u_1 c^\alpha$ one cannot get by simply with the trace of the Bott connection ($\psi = f\xi$) as above: one actually needs the connection and its curvature K. Carrying this out for the trace powers ψ^α instead of the c^α one finds that

$$u_1 = \frac{\text{trace } A}{\|\zeta\|^2} \xi$$

$$\psi^\alpha(K) = \psi^\alpha(A)\|\zeta\|^{-2n} \sum_j (d\bar{\zeta}_0 \wedge dz_0 \wedge \ldots \widehat{d\bar{\zeta}_j \wedge dz_j} \ldots d\bar{\zeta}_n \wedge dz_n) \bmod \xi$$

so that

$$u_1\psi^\alpha(K) = \frac{(\text{trace } A)\ \psi^\alpha(A)}{\|\zeta\|^{2n+2}} \sum (-1)^j \bar{\zeta}_j (d\bar{\zeta}_0 \wedge \ldots \widehat{d\bar{\zeta}_j} \ldots d\bar{\zeta}_n)$$

$$\wedge (dz_0 \wedge \ldots \wedge dz_n) .$$

Now changing entirely into ζ-coordinates one picks up $(\det A)^{-1}$ and integrating over the ζ-unit sphere

$$\int_{S^{2n+1}} u_1 \psi^{\alpha} = \frac{(\text{trace } A)\ \psi^{\alpha}(A)}{\det A}\, c_n$$

where c_n is the volume of S^{2n+1} with respect to the volume form above.

III Relations with the cohomology of formal vector fields

1. Foliations and L_q

We shall show that the characteristic classes constructed above arise naturally as the classes in the cohomology of the Lie algebra of formal vector fields. This comes about in two steps: first we give a global and functorial setting for the local computations with the connection form of a torsion-free connection; and second, we show that the Lie algebra mentioned above gives rise to the "universal" algebra embodying the identities satisfied by the connection and curvature forms of a Bott connection. In this framework the Bott vanishing theorem drops out "automatically", and we could have started with this construction. However, both constructions are useful for computations, and each one enriches the other.

To begin with, we reformulate our notion of foliation somewhat more geometrically. Given an integrable subbundle $E \subset TM$ of codim q, around each point $x \in M$ there is a coordinate chart U such that the leaves of the foliation are the level sets of the last q coordinate functions; this is just Frobenius' theorem. These q functions may be interpreted as giving a smooth submersion $f_U\colon U \to R^q$ so that the leaves are $f_U^{-1}(pt)$ and $E|_U = \ker df_U$. If $f_i\colon U_i \to R^q$ and $f_j\colon U_j \to R^q$ are two such submersions, then in the overlap $U_i \cap U_j$ they are related by a map $\gamma_{ij}\colon U_i \cap U_j \to \Gamma_q =$ pseudo group of germs of diffeomorphisms of open sets in R^q; more precisely, for each $x \in U_i \cap U_j$, $f_i(x) = \gamma_{ij}(x) \circ f_j(x)$. Thus E determines a triple $\{U_i, f_i, \gamma_{ij}\}$. Of course this triple is not unique; what E determines uniquely is a maximal family of such triples $\{U_i, f_i, \gamma_{ij}\}$ where "maximal" is interpreted in the obvious sense. (See Bott [2] or Haefliger [11] for precise details). Conversely, given any open cover $\{U_i'\}$ of M, and submersions $f_i': U_i' \to R^q$ which are related in the overlaps by smooth maps $\gamma_{ij}': U_i' \cap U_j' \to \Gamma_q$, we get an integrable subbundle $E' \subset TM$ of codimension q; over each U_i', $E'|_{U_i'} = \operatorname{Ker} df_i'$ and the local bundles patch together in the overlaps to give a subbundle $E \subset TM$; the integrability follows since if $X, Y \in \Gamma(E|_{U_i})$ then $df_i[X,Y] = X(df_i(Y)) - Y(df_i(X)) - ddf_i(X,Y) = 0$, so $[X,Y] \in \Gamma(E|_{U_i})$. Note in particular that the differentials $d\gamma_{ij}: U_i \cap U_j \to GL_q(R)$ are the transition functions for the normal-bundle $Q = TM/E$, with respect to the trivialization given by $\{U_i, f_i\}$.

Now consider the principal $GL_q(R)$ bundle $\pi: P(Q) \to M$. On $P(Q)$ we have q canonical one-forms $\omega_1, \ldots, \omega_q$ corresponding to the canonical trivialization of $\pi^*(Q)$. These forms span $\pi^*(Q^*)$ and they still generate a differential ideal in $A^*(P(Q))$ (because d commutes with π^*); hence there are forms $\phi_{ij} \in A^1(P(Q))$ such that $d\omega_i = \sum \phi_{ij} \wedge \omega_j$. Of course the ϕ_{ij} are not unique, or even canonical. To get around this point we pass to another bundle over $P(Q)$, where "second order derivatives" will make sense. Precisely, let $P_k(Q)$ = bundle on M whose fibre over $x \in M$ is the space of k-jets of all local submersions $f_U: U \to \underset{\sim}{R}^q$ associated to the foliation which take x to 0. Thus $P_1(Q) = P(Q)$

and each $P_{k+1}(Q)$ is a bundle over $P_k(Q)$ with fiber a cell (for $k \geq 1$). We have then a tower of bundles

$$\longrightarrow P_{k+1}(Q) \xrightarrow{\pi_k} P_k(Q) \longrightarrow \cdots \longrightarrow P_1(Q) \xrightarrow{\pi} M$$

and on these $P_k(Q)$ we will be able to construct canonical connection forms and their derivatives.

To begin with we reconstruct the canonical forms $\omega_1, \ldots, \omega_q$ interpreting $P(Q)$ as $P_1(Q)$. A point on $P(Q)$ is given by $J^1(f) = (x:\partial f/\partial x_1, \ldots, \partial f/\partial x_n)$ where $x \in M$, and $f: U \to \underset{\sim}{R}^q$ is a submersion taking x to 0. A tangent vector X on $P(Q)$ at $J^1(f)$ is specified by a curve starting at $J^1(f)$, i.e., by a 1-parameter family $J^1(f_t)$ of one-jets of $f_t: U \to \underset{\sim}{R}^q$. Thus

$$\left.\frac{\partial f_t}{\partial t}\right|_{t=0} = \frac{\partial f}{\partial x_j} \cdot \frac{\partial x_j}{\partial t}$$

and we see that $\omega_j(X) = \partial x_j/\partial t = j^{th}$ component of $\pi_*(X)$ in the frame $(\partial f/\partial x_1, \ldots, \partial f/\partial x_n)$. This description is of course canonical, and coincides with the one familiar from topology. Now a tangent vector Y on $P_2(Q)$ at $J^2(f)$ is again given by the two-jets of a 1-parameter family, $f_t: U \to \underset{\sim}{R}^q$ and we set

$$\theta_{ij}(Y) = \frac{\partial}{\partial x_j}\left(\frac{\partial x_i}{\partial t}\right)$$

It is a matter of calculus now to show that on $P_2(Q)$, $d\omega_i = \sum \theta_{ij} \wedge \omega_j$ (suppressing the π_2^*). Similarly on $P_3(Q)$ we have q^3 canonical forms $\theta_{ijk} \in A^*(P_3(Q))$ which satisfy $d\theta_{ij} = \sum \theta_{ijk} \wedge \omega_k + \sum \theta_{ik} \wedge \theta_{kj}$ and in general on $P_\ell(Q)$ we have q^{ℓ} canonical one-forms satisfying the appropriate differential identity: see Kobayashi [14] for details. Thus if we pass to the space $P_{\infty}(Q) = \varprojlim_{k} P_k(Q)$ we will have canonical one-forms satisfying the local identities associated to a torsion-free connection on M.

We now seek a universal algebra in which these identities are satisfied. With any luck, the obvious map from this universal algebra into the forms on $P_{\infty}(Q)$ should yield the characteristic classes of the foliation. This will take a little bit of doing, but the universal algebra is easy to describe.

Consider the ring of formal power-series on R^q, $R[[x_1,\dots,x_q]]$, and let L_q denote the free module on q generators $\partial/\partial x_1,\dots,\partial/\partial x_q$ over this ring. L_q can be made into a Lie algebra by the obvious bracket relation

$$\left[\sum f_i \frac{\partial}{\partial x_i}, \sum g_i \frac{\partial}{\partial x_j}\right] = \sum_j \left(\sum_i f_i \frac{\partial g_j}{\partial x_i} - g_i \frac{\partial f_j}{\partial x_i}\right) \frac{\partial}{\partial x_j}$$

where $f_i, g_i \in R[[x_1,\dots,x_q]]$. The ring of formal power-series has a natural M-adic topology where M is the unique maximal ideal of all power series with zero constant term. L_q inherits a topology from this ring, and the bracket operation is continuous. This is the (topological) Lie algebra of formal vector fields on R^q. In a suitable sense, L_q is "the Lie algebra of the pseudo-group Γ_q"; we shall not use this interpretation.

Now L_q can be considered as the direct product of subspaces L_q^k, $-1 \leq k \leq \infty$, where $L_q^{-1} = \left\{\frac{\partial}{\partial x_i}\right\}$, $L_q^0 = \left\{x_i \frac{\partial}{\partial x_j}\right\}$, $L_q^1 = \left\{x_i x_j \frac{\partial}{\partial x_k}\right\}$, ... and the indexing has been so chosen that

$[L_q^k, L_q^\ell] \subset L_q^{k+\ell}$. In particular, L_q^{-1} is an abelian subalgebra of dimension q (corresponding to the subgroup of translations in Γ_q), L_q^0 is the subalgebra $\underline{gl}_q$ of $GL_q \subset \Gamma_q$. Note that L_q^0 acts on each subspace L_q^k -- indeed this is just the action of GL_q on the space $S^k(R^q)^* \otimes R^q$, where $S^k(\cdot) = k^{th}$ symmetric power.

To get the universal algebra we are after, we look at the exterior algebra of the M-adically continuous dual of L_q. Let $L_q^* = \{f: L_q \to R | f$ is linear, and continuous in the M-adic topology$\}$. Then it is easy to see that L_q^* is the direct sum of the linear duals of the L_q^k, and hence has a basis $\{\omega_i, \theta_{ij}, \ldots\}$ with ω_i spanning the dual of L_q^{-1}, $\{\theta_{ij}\}$ spanning the dual of $L_q^0, \ldots$. In the exterior algebra $\Lambda_R^*(L_q^*)$ we can define a differential d dual to the bracket, and we find *

$$d\omega_i = \sum \theta_{ij} \wedge \omega_j , \quad d\theta_{ij} = \sum \theta_{ijk} \wedge \omega_k + \sum \theta_{ik} \wedge \theta_{kj}, \ \ldots \ .$$

These are precisely the relations we are after. Thus if we define $A^p(P_\infty(Q)) = \lim\limits_{\substack{\to \\ n}} A^p(P_n(Q))$ then there is an obvious map $\Phi: \Lambda_R^*(L_q^*) \to A^*(P_\infty(Q))$ which is specified by requiring that the universal ω_i map to the canonical ω_i on $P_\infty(Q)$, the universal θ_{ij} map to the canonical θ_{ij} on $P_\infty(Q)$ and so on. The map commutes with d by construction, and so we have a map on cohomology

*The forms $d\theta_{ijk}$ and further on are not quite the dual forms to the basis for L_q^k described above: they need to be modified by appropriate binomial coefficients to give the differential relations as stated. See [11].

$$\Phi^*: H^*(L_q^*) \to H^*_{DR}(P_\infty(Q))$$

where the right-hand side is by definition the cohomology of $A^*(P_\infty(Q))$.

We are still a little ways away from our goal of getting a map to $H^*(M)$; but we can begin to pare Φ^* down at once. Recall first that direct limits commute with homology of (cochain) complexes, and so $H^*_{DR}(P_\infty(Q)) = \varinjlim_n H^*_{DR}(P_n(Q))$. Now for $n \geq 1$, each $\pi_n: P_{n+1}(Q) \to P_n(Q)$ is a homotopy equivalence, so $H^*_{DR}(P_\infty(Q)) \simeq H^*_{DR}(P_1(Q))$. Hence we have

$$\Phi^*: H^*(L_q) \to H^*(P_1(Q)) \ .$$

In the special case that Q happens to be trivial, we are essentially done. For then we can choose a section $\sigma: M \to P_1(Q)$ and we get $\Phi^*: H^*(L_q) \to H^*(P_1(Q)) \xrightarrow{\sigma^*} H^*(M)$. We now need to know of course that the cohomology of L_q is; the general result is deferred to the next section but we shall work out the special case of $q = 1$ below. Note that the map σ^* on cohomology is unique, up to choosing an orientation for the trivial bundle Q.

Let us now attack the more general case in which Q is not required to be trivial. $GL_q(R)$ acts freely on $P_1(Q)$ and $P_1(Q)/GL_q \simeq M$; in fact up to homotopy we may replace $GL_q(R)$ by its maximal compact subgroup O_q , since $GL(R)/O_q$ is diffeomorphic to a Euclidean space. Then $P_1(Q)/O_q \sim M$; moreover, the forms on $P_1(Q)$ which correspond to forms on M are precisely the "O_q-basic" ones (see Cartan [5] or Haefliger [11]). Now O_q acts on all the $P_n(Q)$, $1 \leq n \leq \infty$, and we may therefore consider the

O_q basic ones; O_q also acts on L_q (recall $o_q \subseteq \underline{gl}_q \subseteq L_q$) and thus we have a map

$$\Phi' : \Lambda^*(L_q, o_q) \to A^*(P_\infty(Q)/O_q) = \lim_{\vec{n}} A^*(P_n(Q)/O_q)$$

induced from Φ, which is compatible with the O_q-actions. As before, we have a map on cohomology

$$\Phi'^* : H^*(L_q, o_q) \to H^*_{DR}(P_\infty(Q)/O_q)$$

and again

$$H^*_{DR}(P_\infty(Q)/O_q) = \lim_{\vec{n}} H^*(P_n(Q)/O_q) \simeq H^*(P_1(Q)/O_q) \simeq H^*(M) .$$

Thus finally we have a commutative diagram

$$\begin{array}{ccc} H^*(L_q) & \xrightarrow{\Phi^*} & H^*(P_1(Q)) \\ \uparrow & & \uparrow \pi^* \\ H^*(L_q, o_q) & \xrightarrow{\Phi'^*} & H^*(M) \end{array}$$

and Φ'^* factors through Φ^* when $\pi : P_1(Q) \to M$ has a section.

In the next section we shall see that $H^*(L_q, o_q)$ is canonically isomorphic with the cohomology of WO_q, thus giving the promised construction of characteristic classes. To end, we give a simple illustration which shows that $H^*(L_1)$ captures the Godbillon-Vey class and in fact nothing more.

Example. We want to compute the cohomology of L_1 and relate it to the Godbillon-Vey class of a codimension 1 foliation. L_1 is the direct product of the one-dimensional subspaces generated by $Y_k = x^{k+1} \frac{d}{dx}$, $-1 \leq k$; and the bracket is obviously $[Y_i, Y_j] = (j-i)Y_{i+j}$. Thus if we let $\alpha_{-1}, \alpha_0, \alpha_1, \ldots$ be the basis for L_1^* dual to the Y_i's we see at once that $d\alpha_k = \sum_{i<k-i} (k-2i)\alpha_i \wedge \alpha_{k-i}$ in $\Lambda^2(L_q^*)$. The direct computation of the d in the higher exterior powers gets complicated, but we can reduce the problem considerably by considering the adjoint action of Y_0 (which generates $\underline{gl}_1$) on $\Lambda^*(L_1^*)$.

The bracket relation $[Y_0, Y_j] = jY_j$ shows that each α_j is an eigenvector for $\mathrm{ad}(Y_0)$ of eigenvalue j, and hence that $\alpha_{j_1} \wedge \ldots \wedge \alpha_{j_p}$ is an eigenvector for $\mathrm{ad}(Y_0)$ of eigenvalue $\sum_{i=1}^{n} j_i$. Now decompose $\Lambda^*(L_1^*)$ into the eigenspaces of $\mathrm{ad}\ (Y_0)$: $\Lambda^*(L_1^*) = \sum_{r=-1}^{\infty} C_r^*$, C_r^* being the graded subspace of elements which are of eigenvalue r. As d commutes with $\mathrm{ad}(Y_0)$ * each C_r^* is a subcomplex of $\Lambda^*(L_1^*)$; and we claim that for $r \neq 0$, C_r^* is acyclic. Indeed, if $\beta \in C_r^*$ is a cocycle and $r \neq 0$, $r\beta = \mathrm{ad}(Y_0)\beta = di_{Y_0}\beta$; and one checks trivially (because i_{Y_0} commutes with $\mathrm{ad}(Y_0)$)* that $i_{Y_0}\beta \in C_r^*$. Thus we have shown that $H^*(\Lambda^*(L_1^*)) \simeq H^*(C_0^*)$.

Now a monomial $\alpha_{j_1} \wedge \ldots \wedge \alpha_{j_p}$ can have weight (= eigenvalue) zero only if $p \leq 3$: for at most one of the j_i's is negative, and then it is -1. C_0^* is thus zero in degrees $p \geq 4$ and is generated by 1, $C_0^1 = \{\alpha_0\}$, $C_0^2 = \{\alpha_{-1} \wedge \alpha_1\}$ and $C_0^3 = \{\alpha_{-1} \wedge \alpha_0 \wedge \alpha_1\}$. The formula $d\alpha_0 = 2\alpha_{-1} \wedge \alpha_1$ shows that α_0 is not a cocycle, and $\alpha_{-1} \wedge \alpha_1$ is a coboundary. Thus it follows that $H^*(L_1) \simeq H^*(C_0^*) = \Lambda(y)$, deg y = 3, y represented by the form $\alpha_{-1} \wedge \alpha_0 \wedge \alpha_1$.

It comes to essentially the same computation to compute $H^*(L_1)$

*Use the relation $\mathrm{ad}(Y) = i_Y d + di_Y$, given in the Appendix.

from the spectral sequence for the pair $(L_1, \underline{gl_1})$: clearly $H^*(\underline{gl_1}) = \Lambda(\alpha_0)$ and the above argument is more or less equivalent to $H^*(L_1, \underline{gl_1}) \simeq \Lambda(x)$, deg $x = 2$, represented by $\alpha_{-1} \wedge \alpha_1$: now $d_2\alpha_0 = 2x$, whence the result.

To relate this to the Godbillon-Vey class, let $E \subset TM$ be a codimension 1 foliation with normal bundle $Q = TM/E$. On $P_1(Q)$ we have a canonical 1-form ω, on $P_2(Q)$ a canonical θ satisfying $d\omega = \theta \wedge \omega$, and on $P_3(Q)$ a θ_1 satisfying $d\theta = \theta_1 \wedge \omega$, and so on a canonical θ_k on each $P_k(Q)$. Hence the map $\Phi: L_1^* \to A'(P_\infty(Q))$ is given by $\Phi(\alpha_{-1}) = \omega$, $\Phi(\alpha_0) = \theta$, $\Phi(\alpha_k) = \theta_k$, $k \geq 1$. On cohomology we have $\Phi^*: H^*(L_1) \to H^*(P_\infty(Q)) \simeq H^*(P_1(Q))$ given by $\Phi^*(\alpha_{-1} \wedge \alpha_0 \wedge \alpha_1) = \{\omega \wedge \theta \wedge \theta_1\} = \{\theta \wedge d\theta\}$. When Q is trivial, this is exactly the Godbillon-Vey class discussed earlier: if not, it gives the Godbillon-Vey class on a double-cover, which in fact is $O(1)$-basic, so descends to $H^*(M)$.

In this set-up one can give a quick and easy computation of Roussarie's examples, and of <u>one</u> of Bott's family. First consider the algebra.

If we consider the generators $H = \begin{pmatrix} 1 & 0 \\ 0 & -1 \end{pmatrix}$, $X = \begin{pmatrix} 0 & 1 \\ 0 & 0 \end{pmatrix}$, $Y = \begin{pmatrix} 0 & 0 \\ 1 & 0 \end{pmatrix}$ for $sl(2,R)$, there is a Lie algebra homomorphism $\phi: sl(2,R) \to L_1$ given by $\phi(H) = 2Y_0$, $\phi(X) = Y_1$, $\phi(Y) = Y_{-1}$. The dual map $\phi^*: L_1^* \to sl(2,R)^*$ is then given by $\phi^*(\alpha_{-1}) = \eta$, $\phi^*(\alpha_0) = 2\theta$, $\phi^*(\alpha_1) = \xi$, $\phi^*(\alpha_k) = 0$ for $k \geq 2$. Hence the map on cohomology $\phi^*: H^*(L_1) \to H^*(sl(2,R))$ is an isomorphism. Complexifying the situation, $\phi \otimes 1: sl(2,C) \to L_1^C = L_1 \otimes C$ gives an isomorphism $\phi^* \otimes 1: H^*(L_1^C) \to H^*(sl(2,C))$.

Geometrically ϕ arises from the action of $SL(2,R)$ on its homogeneous space $S^1 = SL(2,R)/B$, B being the subgroup of upper triangular matrices. This is precisely Roussarie's foliation, at the group level. Now the top (and only) generator of $H^*(sl(2,R))$ is a volume form (by unimodularity -- see Koszul [16] for the general proof) and so descends to a volume form on any compact $M = SL(2,R)/\Gamma$.

For Bott's example, consider the vf $X = z_1 \frac{\partial}{\partial z_1} + z_2 \frac{\partial}{\partial z_2}$. On $C^2-\{0\}$ this is the infinitesimal generator of the C^*-action given by scalar multiplication: hence if we pull up the foliation to $SL(2,C)$, it is given by cosets of the Borel subgroup B^C of upper triangular matrices; hence $\phi \otimes 1$ is the corresponding map. Now $\pi: SL(2,C) \to C^2-\{0\}$ is an isomorphism on cohomology, and $H^*(sl(2,C)) \simeq H^*(SL(2,C):C)$ so the G-V class of the foliation on the group, and hence $C^2-\{0\}$, is not zero.

2. The cohomology of L_q and $(L_q, o(q))$

We describe the main steps in the computation of $H^*(L_q)$ and $H^*(L_q, o(q))$ following the exposé of Godbillon [9], where the reader will find more details. There are essentially two steps in the computation of $H^*(L_q)$: first to construct a map from the truncated Weil algebra of $\underline{gl}_q$ to $\Lambda^*(L_q^*)$, and then to show -- by analyzing the action of $\underline{gl}_q$ on $\Lambda^*(L_q^*)$ (compare last section) -- that this map is an isomorphism on the E_1-level of spectral sequences associated to the two algebras. Once $H^*(L_q)$ is computed, the relative case follows easily.

Before discussing the two steps in more detail, we give a description of what the end result is. There are classes

$u_1,\ldots,u_q$ in $E_1^{0,*}$ (deg $u_i = 2i-1$) and $c_i \in E_1^{2i,0}$, $1 \leq i \leq q$ such that the following monomials form an additive base for $H^*(L_q)$: $u_{i_1} \ldots u_{i_r} c_{j_1} \ldots c_{j_2}$ where the sequences $(i_1,\ldots,i_r)$ and $(j_1,\ldots,j_s)$ satisfy

(i) $1 \leq i_1 < i_2 < \ldots < i_r \leq q$, $1 \leq j_1 \leq \ldots \leq j_s \leq q$,

(ii) $\sum_1^s j_k \leq q$,

(iii) $i_1 + \sum_1^s j_i > q$.

Note that deg $(u_{i_1} \ldots u_{i_r} c_{j_1} \ldots c_{j_s}) = \sum_1^r (2i_k-1) + 2\sum_1^s j_k$. Hence the above results imply

(a) $H^i(L_q) = 0$ for $1 \leq i \leq 2q$, $i \geq q^2+2q$

(b) All products in $H^*(L_q)$ are zero.

The first consequence (a) is quite reasonable; if we regard $H^*(L_q)$ as being the universal algebra for foliations of codim q with trivial normal bundle, it is reasonable to expect that all primary classes (which are in degrees $\leq 2q$) vanish. The second half which says that $H^*(L_q) = 0$ above q^2+2q, can be seen by analysing $\Lambda^*(L_q^*)$ as a module over the <u>centre</u> of $\underline{gl}_q$. The centre is one-dimensional, generated by $H = \sum_1^q x_i \frac{\partial}{\partial x_i}$. The map ad(H) commutes with d, and so we can break up $\Lambda^*(L_q)^* = \sum_\lambda C_\lambda^*$ where C_λ^* is the subcomplex of the elements of eigenvalue λ for ad(H). Exactly as in the previous section, C_λ^* is acyclic if $\lambda \neq 0$, C_0^* is concentrated in degrees $\leq q^2+2q$.

Now let us discuss the full computation of $H^*(L_q)$. There is a natural Lie algebra homomorphism $\pi\colon L_q \to \underline{gl}_q$ obtained by assigning to a formal vector field its linear part: and clearly

π is a retraction for the inclusion $\underline{gl}_q \hookrightarrow L_q$. In the terminology of Cartan [5] (see Appendix I, Section 4) this is a $\underline{gl}_q$-connection and so gives rise to a map $\psi: W(\underline{gl}_q) \to \Lambda^*(L_q^*)$ where $W(\underline{gl}_q) = \Lambda^*(\underline{gl}_q^*) \otimes S^*(\underline{gl}_q^*)$ is the Weil algebra of $\underline{gl}_q$. Recall how ψ is constructed. The map π above gives $\pi^*: \underline{gl}_q^* \to L_q^*$ and hence $\Lambda^*(\pi^*): \Lambda^*(\underline{gl}_q^*) \to \Lambda^*(L_q^*)$. This also leads to a map $K: S^*(\underline{gl}_q^*) \to \Lambda^{ev}(L_q^*)$ as follows: on $S^1(\underline{gl}_q^*) = \underline{gl}_q^*$, $K(\omega) = d\pi^*(\omega) - \Lambda^2(\pi^*)\, d\omega$, or symbolically $K = [d,\pi^*]$, and then one extends K multiplicatively. The map ψ is then the composition given by the following diagram:

$$W(\underline{gl}_q) \xrightarrow{\Lambda^*(\pi^*)\ \otimes\ K} \Lambda^*(L_q^*) \otimes \Lambda^{ev}(L_q^*) \xrightarrow{\text{mult}} \Lambda^*(L_q^*) .$$

In the case at hand, we shall show that ψ is zero on the ideal J_q in $W(L_q)$ generated by $S^{q+i}(L_q^*)$, $i > 0$, and thus factors through to give $\bar{\psi}: W_q = W(\underline{gl}_q)/J_q \to \Lambda^*(L_q^*)$. Suitably interpreted, ψ is the "universal Bott connection" and the above statement is then precisely Bott's vanishing theorem. So let us look more closely at ψ. Writing out $L_q^* = \sum_{k=-1}^{\infty} L_q^{k*}$, we see that $\pi^*: \underline{gl}_q^* \to L_q^*$ is given by $\pi^*(\theta_{ij}) = \theta_{ij}$.† Hence $K(\theta_{ij}) = d_{L_q}(\theta_{ij}) - \Lambda^2(\pi^*)\, d_{\underline{gl}_q}(\theta_{ij}) = \sum_{k=1}^{q} \theta_{ijk} \wedge \theta_k$. It now follows exactly as in the proof of Bott's theorem that K is zero on any monomial $\theta_{i_1 j_1} \cdots \theta_{i_r j_r}$ for $r > q$.

Now W_q, $\Lambda^*(L_q)$ are filtered DGA's, and $\bar{\psi}$ is a map preserving this structure. We shall see that on the associated spectral sequences (which abut to the cohomology of the respective DGA's) $\bar{\psi}$ is an isomorphism on the E_1-level, and so $\bar{\psi}: H^*(W_q) \xrightarrow{\sim} H^*(L_q)$.

†par "abus de langage".

In the course of proving this isomorphism, we shall come up with the classes u_i and c_i mentioned earlier. The final point in the computation is to sit down and work out the nitty-gritty in $H^*(W_q)$; as the reader must have suspected, we shall omit this.

Let's define the filtrations and see what the E_1-terms are. The foliation on W_q is simply Cartan's filtration [5] on $W(\underline{gl}_q)$ "pushed down": explicitly, $F^p W_q^r = \{w \in W_q^{p+r} : i_{X_1} \dots i_{X_{r+1}} w = 0$ for all $X_1, \dots, X_{r+1} \in \underline{gl}_q\}$. Exactly as in Cartan (loc. cit.) the E_1-term is then $H^*(\underline{gl}_q, S_q^r(\underline{gl}_q^*)) \simeq H^*(\underline{gl}_q) \otimes I_S(\underline{gl}_q)$, the second factor of course being truncated above degree q. The u_i are then the primitive generators of the first term, the c_i of the second. When one deals with the spectral sequence for the full Weil algebra, the E_∞-term is zero because the algebra is acyclic: thus one finds that the u_i are obtained from the c_i by "transgression". The recipe for the transgression map at the level of forms is to carry out a formal "integration over the fibre" as discussed in Chapter II, Section 4, Example b.

The filtration on $\Lambda^*(L_q^*)$ is the Hochschild-Serre filtration relative to the subalgebra $\underline{gl}_q$. A glance at H-S [13] shows that the E_1-term is then $H^*(\underline{gl}_q) \otimes B^*$, where $B^* = \{\phi \in \Lambda^*(L_q^*)$: $i_X = L_X = 0$ for all $x \in \underline{gl}_q\}$. Using the fact that $L_q^{k*} = S^k(R^{q*}) \otimes R^q$ as $\underline{gl}_q$ modules one easily identifies this cohomology with the E_1-term above, noting that $\bar{\psi}$ gives the isomorphism.

The inclusions $o(q) \subseteq \underline{gl}_q \subseteq L_q$ give actions of $o(q)$ on $\Lambda^*(L_q^*)$ and W_q, and these are respected by the map $W_q \to \Lambda^*(L_q^*)$. Hence one can pass to the subalgebras $o(q)$-basic elements, and get a map $\bar{\psi}: WO_q \to \Lambda^*(L_q^*, o(q)) = \{\phi \in \Lambda^*(L_q^*): i_X\phi = L_X\phi = 0$ for $X \in o(q)\}$. The filtrations restrict to these subalgebras and $\bar{\psi}$ respects them, and one checks again that $\bar{\psi}$ is an isomorphism on the E_1-level.

It remains to describe WO_q in terms of the object of the same name introduced in Chapter II, and to describe its cohomology. Of course WO_q is precisely the algebra $\Lambda(u_1, u_3, \ldots, u_{2\ +1}) \otimes R[c_1, \ldots, c_q]/(\text{terms of degree} > q)$ introduced in Chapter II. To see this consider the $o(q)$-basic elements in $\Lambda^*(\underline{gl}_q^*)$. These are precisely $\Lambda^*((\underline{gl}_q/o(q)^*)^{o(q)}$. Now as an $o(q)$-module, $(\underline{gl}_q/o(q))^*$ can be identified with the space of symmetric matrices, with $O(q)$ acting by the adjoint action. The invariant subspace in the symmetric matrices is just the scalar multiples of the identity, and the invariants in the higher exterior powers can be computed from the basic theorem on invariants of $o(q)$ to give the resulting exterior algebra. A similar argument identifies the symmetric part. The cohomology of WO_q has been computed by J. Vey; the generators are again monomials $u_I\ c_J$ satisfying the following relations:

(1) $\sum jk \leq q$,

(2) $i^0 + \sum jk > q$,

(3) $i^0 \leq j^0$.

Here $i^0 = \min\{i_1, \ldots, i_m\}$ = min element of I,

j^0 = min (odd indices in J) .

3. L_q and distributions

In section 1 we introduced the lie algebra L_q and the maps Φ and Φ' in a purely formal and algebraic way. We shall now say a few words about the geometric underpinnings of these objects. The "local" construction over $\underset{\sim}{R}^q$ will give the geometric interpretation of L_q , and the corresponding "global" one over arbitrary M will then give Φ and Φ'.

Let A_q be the lie algebra of all C^∞ vector fields on $\underset{\sim}{R}^q$: thus $A_q = \left\{\sum_1^q f_j \frac{\partial}{\partial x_j} : f_j \in C^\infty(\underset{\sim}{R}^q)\right\}$. As a linear space A_q is just the direct sum of q copies of $C^\infty(\underset{\sim}{R}^q)$, and so it inherits the C^∞-topology from $C^\infty(\underset{\sim}{R}^q)$: the lie bracket is easily seen to be continuous in this topology, and so A_q is a topological lie algebra. We may therefore consider its continuous (Gelfand-Fuks) cohomology. Specifically, let $A_q^* = \left\{\phi: A_q \to \underset{\sim}{R} \mid \phi \text{ is linear and continuous in the } C^\infty\text{-topology}\right\}$. Since the bracket is continuous, for each $p \geq 0$ we can define $d: \Lambda^p(A_q^*) \to \Lambda^{p+1}(A_q^*)$ by requiring it to be a derivation and stipulating that on $\Lambda^1(A_q^*)$ it be given by the usual formula $d\phi(X,Y) = -\phi([X,Y])$. Now A_q^* is just q-copies of the Schwartz space of compactly supported distributions on $\underset{\sim}{R}^q$, and so one can meaningfully speak of the support of an arbitrary element $\psi \in \Lambda^p(A_q^*)$. One easily checks then that supp $(d\phi) \subseteq$ supp (ϕ) and so for every subset $S \subseteq \underset{\sim}{R}^q$ the subset $\Lambda_S^*(A_q^*)$ of elements with support in S is a sub-differential-graded algebra. We shall be interested in the subcomplex of those "distributional forms" which are supported at the origin: $\Lambda_0^*(A_q^*)$.

Let J^k_0 be the space of k-jets at 0 of vector fields in A_q: for each k we have a continuous linear map $\pi_k: A_q \to J^k_0$ assigning to a vector field its k-jet at 0. J^k_0 is of course not a lie algebra because the bracket of k-jets depends upon the (k+1)st jet. However, the formal ∞-jets at 0, namely $\varprojlim J^k_0$ do form a topological lie algebra, with the usual inverse limit topology. In fact it is trivial to see that $\varprojlim J^k_0 \simeq L_q$ as a topological lie algebra, and that $\pi = \varprojlim \pi_k: A_q \to L_q$ is a continuous lie homomorphism. Thus we have an induced map $\pi^*: L^*_q \to A^*_q$ on the (continuous) duals. Because π assigns to $X \in A_q$ its formal jet at 0, it is easy to see that $\text{Im}\ \pi^*$ $\subseteq$ elements of A^*_q supported at 0. Conversely every element ω of A^*_q supported at 0 may be interpreted as a continuous linear functional on L_q, since in fact ω only depends on a finite jet of X. Thus $\pi^*: L^*_q \to A^*_q$ (supp 0) is an isomorphism, and this extends to an isomorphism of DGA's, $\pi^*(\Lambda^*(L^*_q), d) \to (\Lambda^*_0(A^*_q), d)$. ($\pi^*$ commutes with d because π is a lie homomorphism). Here then we finally have a geometric representation of L_q and its complex $(\Lambda^*(L^*_q), d)$.

In the case q = 1, one can write out π^* explicitly as follows. Recall that the subspace of A^*_1 supported at 0 is spanned by the Dirac delta function δ and its (distributional) derivatives. To write the differential in the complex $\Lambda^*_0(A^*_1)$ let us agree to write δ as δ_0, and to define δ_k by $\delta_k(f) = (-1)^k \left.\frac{d^k f}{dx^k}\right|_{x=0}$. Then one finds

$$d\delta_k = \sum_{j=0}^{[\frac{k+1}{2}]} \frac{k-2j+1}{k-j+1} \binom{k}{j} \delta_j \wedge \delta_{k-j+1}$$

For k = 0, we work this out from the definition:

$$d\delta(f\frac{d}{dx}, g\frac{d}{dx}) = -\delta[f\frac{d}{dx}, g\frac{d}{dx}] = \delta(f'g) - \delta(fg')$$

$$= f'(0)g(0) - g'(0)f(0) = \delta \wedge \delta_1(f\frac{d}{dx}, g\frac{d}{dx})$$

For $k > 0$ this brute force method gets to be unpleasant, and it is easier to observe that $d\delta_k$ can be calculated by the following inductive procedure. The shift operator $T: \Lambda_0^1(A_1^*) \to \Lambda_0^1(A_1^*)$ given by $T\delta_k = \delta_{k+1}$ is precisely the lie-derivative L_Y where $y = d/dx \in A_1$ is the axial vector field: $L_Y(\delta_k)(f\frac{d}{dx}) = -\delta_k[Y, f\frac{d}{dx}] = -\delta_k(f') = \delta_{k+1}(f\frac{d}{dx})$. Now d commutes with all lie derivatives and so $d\delta_{k+1} = d(L_Y\delta_k) = L_Y(d\delta_k)$.

<u>Proposition</u>. The map $\pi: A_1 \to L_1$ induces an isomorphism of cochain complexes

$$\pi^*: (\Lambda^*(L_1^*), d) \to (\Lambda_0^*(A_1^*), d) .$$

<u>Proof</u>: In terms of the basis $\{\alpha_j \mid j \geq -1\}$ of L_1^* used in section 1, define $\beta_j = (j+1)!\,\alpha_j$. The β_j form a basis dual to the elements $\frac{1}{(j+1)!} x^{j+1} \frac{d}{dx} \in L_1$. Thus the map of exterior algebras induced by $\beta_j \mapsto \delta_{j+1}$ is an isomorphism of algebras. To see that it commutes with the differentials, let $Y_{-1} = \frac{d}{dx} \in L_1$ and note that $L_{Y_{-1}}(\beta_j) = \beta_{j+1}$: hence it suffices to check that $d\delta_{-1}$ goes to $d\delta$, and that $L_{Y_{-1}}$ is transformed into L_Y. We leave it to the reader to notice that this is precisely the map π^*. □

Without entering into all the details, we indicate briefly how one can make the isomorphism π^* explicit for $q > 1$. Now A_q is q copies of $C^\infty(\underset{\sim}{R}^q)$, hence there are q distinct delta functions $\delta^1,\ldots,\delta^q$ in $\Lambda_0^*(A_q^*)$ given by projection in the jth factor composed with δ. As before $\Lambda_0^1(A_q^*)$ is spanned by $\{\delta^i, D^\alpha\delta^i\}$ where $D^\alpha = \partial^{\alpha_1}/\partial x_1^{\alpha_1} \ldots \partial^{\alpha_q}/\partial x_q^{\alpha_q}$. Recall that a basis of $\Lambda^1(L_q^*)$ is given by $\{\omega_i, \theta_{ij}, \theta_{ijk}, \ldots\}$. $\pi^*(\delta^i) = \omega_i$, $\pi^*(-\frac{\partial}{\partial x_j}\delta^i) = \theta_{ij}$ with the higher derivatives involving cascades of binomial coefficients.

We now repeat the same construction in a global setting -- that is, on a manifold M with a codimension q integrable subbundle $E \subseteq TM$ -- to get a canonical map $\Phi: L_q^* \to A^1(P_\infty(E))$. Recall that given an integrable subbundle $E \subseteq TM$, in section 1 we constructed a sequence of principal bundles $\pi_k: P_k(E) \to M$, $k \geq 1$; the fibre of $P_k(E)$ over $x \in M$ is the space of germs of k-jets at 0 of local submersions (around x) defining E. It follows that for any (smooth) section S of $\pi_k^*(Q) = \pi_k^*(TM/E)$ over $P_k(E)$ the k-jet at 0 of S is well defined as a map $\Gamma_{C^\infty}(\pi_k^*(Q)) \to J_0^k$. Hence for each k there is a map $\Gamma_{C^\infty}(T(P_k(E))) \to J_0^k$ since the tangent bundle of $P_k(E)$ naturally projects onto $\pi_k^*(TM)$ and thence to $\pi_k^*(Q)$. Taking inverse limits over k we get a map from the lie algebra of smooth vector fields on P (E), call it X, to L_q; and hence by dualizing $\Phi: L_q^* \to A^1(P_\infty(E))$. Φ extends naturally to the exterior algebras $\Phi: \Lambda^*(L_q^*) \to A^*(P_\infty(E))$ and this map commutes with d because it is the dual of a continuous lie algebra homomorphism $X \to L_q$.

It is a simple matter to check that this gives the same map as the one constructed in section 1. Notice first that the construction of Φ just given is natural with respect to smooth maps transverse to E: in particular it may be restricted to open sets of M. If we choose an open cover over which Q is trivial, then all the bundles $P_k(E)$ are trivial as well, and our construction reduces essentially to the map $(\Lambda^*(L_q^*), d) \rightarrow (\Lambda_0^*(A_q), d)$ constructed earlier: thus locally the two constructions coincide, and hence globally as well. Alternatively, we remark that each $\pi_k^*(Q)$ has a canonical trivialization, and the jet maps $\Gamma(\pi_k^*(Q)) \rightarrow J_0^k$ are precisely dual to the canonical one-forms on $P_k(E)$ which give the exterior derivatives of the canonical basis for $\Gamma(\pi_k^*(Q^*))$.

Now one defines the maps Φ^* and Φ'^* on cohomology in exactly the same way as in section 1. The map Φ induces a map on cohomology

$$\Phi^*: H^*(L_q) \rightarrow H^*_{DR}(P_\infty(E)) \simeq H^*(P_1(E))$$

and now depending on whether Q is trivial or not, one has either
$\Phi^{*'}: H^*(L_q) \rightarrow H^*(M)$ (if $P_1(E)$ admits a section) or
$\Phi^*: H^*(L_q, o(q)) \rightarrow H^*(P_1(E)/O(q)) \simeq H^*(M)$.

One can view this construction as a "refinement" of the classical Chern-Weil procedure for defining characteristic classes; it gives the primary and secondary classes at one fell swoop, and it is entirely canonical and free of the introduction of auxiliary objects such as connections. Of course, it only works for the normal-bundle of a foliation; this includes TM (for the foliation by points) but it does not include all bundles on M.

To end we remark that we identified the complexes $(\Lambda^*(L_q^*), d)$ and $(\Lambda_0^*(A_q^*), d)$ and hence by section 1 we have a computation of $H^*(\Lambda_0(A_q^*), d)$ = Gelfand-Fuks of $\mathbb{R}^q$ supported at 0. One can also ask for a computation of the cohomology of the full distributional complex $(\Lambda^*(A_q^*), d)$: in fact the inclusion $\Lambda_0^*(A_q^*) \rightarrow \Lambda^*(A_q^*)$ induces an isomorphism in cohomology. We shall not pursue this here as it really leads into other topics.

IV Variation of secondary characteristic classes

1. Variation of foliations

In the setting of the cohomology of formal vector fields, we shall now discuss the possible variation -- or more precisely, first-order variation -- of secondary classes. As we have seen in the examples of Bott and Thurston, the secondary classes can vary smoothly with parameters and we now set up a way of detecting this variation to first order (i.e., infinitesimally) following Gelfand, Feigin and Fuks [8]. We define the first order variation of a foliation and define characteristic classes for this object. We also define a variation homomorphism which gives essentially the derivative (along the parameters) of the characteristic classes. In this way we shall recapture the result of Simons [19] and Heitsch [12] that for a foliation of codimension q, all classes in the image of $H^*(L_{q+1}) \to H^*(L_q)$ are rigid. We should point out that Gelfand et al. [8] only discuss one-parameter variation but the more general case follows easily from their blueprint.

Let us begin by defining what we mean by an r-parameter variation of a foliation.* So consider $E \subseteq TM$, a foliation of codimension q. Let T be an auxiliary r-dimensional manifold,

*What we call "variation" is also called "deformation": we have adhered to the terminology of [8] which is preferable in that at the level of Lie algebras, the term deformation is misleading, see Section 2.

which will be the space of parameters, and in practice it will of course be R^r or some open set in it.

<u>Definition</u>. A variation of $E \subseteq TM$ with parameters T is an integrable subbundle $\mathcal{E} \subseteq T(M \times T)$ of <u>codimension</u> $q+r$ satisfying

(i) for every $t \in T$, $\mathcal{E}_t \overset{\text{def}}{=} \mathcal{E}|_{M\times\{t\}}$ is an integrable subbundle of codimension q in $T(M \times \{t\})$,

(ii) $\mathcal{E}_0 = E$, where $\mathcal{E}_0 = \mathcal{E}_{t_0}$ for some base-point $t_0 \in T$.

Thus $\mathcal{E}$ is essentially a varying family $E_t \subseteq TM$ of integrable subbundles of codim q. Note that even when $T = R$ or $[0,1]$ this is quite different from an integrable homotopy -- for $\mathcal{E}$ has codimension $q+r$, not q, in $T(M \times T)$. In fact, every integrable homotopy is a variation of a restrictive sort -- but of course, not conversely. For if F is an integrable homotopy of E_0, E_1 say on $M \times [0,1]$ then at each $M \times [t]$, the intersections $F \cap T(M \times \{t\}) = \mathcal{E}_t$ fit together to give a variation. In order to define the equivalence up to first order of two such families, we reformulate the definition.

Consider a variation $\mathcal{E} \subseteq T(M \times T)$ of E. We can find a coordinate covering $\{U_\alpha\}$ of $M \times T$ and submersions $f_\alpha\colon U_\alpha \to R^q \times R^r$ (given by the "last q" coordinates in M and all coordinates in T) such that $\mathcal{E}|_{U_\alpha} = \text{Ker}\, df_\alpha$. Moreover, the following diagram commutes.

$$\begin{array}{ccc} U_\alpha & \xrightarrow{f_\alpha} & R^q \times R^r \\ & \searrow^{\pi} \quad \swarrow^{\pi_r} & \\ & R^r & \end{array}$$

where π_r and π are induced from projections in the second factor. In the overlaps $U_\alpha \cap U_\beta$, f_α and f_β are related by $\Gamma_{\alpha\beta} \in \Gamma^r_{q+r}$ = group of germs of diffeomorphisms of open sets in $R^q \times R^r$ which are the identity on the second factor. Finally $\mathcal{E}_0 = E$ says that at the base point to this data reduces to the corresponding data for E. It is clear that a maximal "chart" of this form is equivalent to giving $\mathcal{E}$.

In coordinates, what the above conditions boil down to is this: f_α is completely specified by q functions: $f_\alpha(x,t) = (f^1_\alpha(x,t), \ldots, f^q_\alpha(x,t), t)$ where by "abus de langage" we denote by t a point in T and its coordinates in U_α. The maps $\gamma_{\alpha\beta}$ are then given by a $(q+r) \times (q+r)$ matrix of functions of the form $\left(\begin{array}{c|c} * & * \\ \hline 0 & I_r \end{array}\right)$ with the obvious (q,r) partitioning.

With this understood we say that two variations $\mathcal{E}$ and $\mathcal{E}'$, with parameters T are equivalent to first order if and only if for every sufficiently fine open covering of $M \times T$ the 1-jets along T of f_α, f'_α taking (x,t_0) to 0 agree. In particular, $f_\alpha(x,t_0) = f'_\alpha(x,t_0)$ so that $\mathcal{E} = \mathcal{E}'_0 = E$. For such equivalence classes we shall now define secondary characteristic classes.

Let us now look at the jet bundles associated to $\mathcal{E}$ and single out the differential forms of interest: this will then lead to the appropriate universal algebra. Set $\mathcal{Q}^* = (T(M \times T)/\mathcal{E})^*$, the annihilator bundle of $\mathcal{E}$ and consider the bundles $P_k(\mathcal{Q})$ = bundle whose fibre over (x,t) is the space of k-jets of the submersions f_α taking (x,t) to 0. On $P_1(\mathcal{E})$ we have q+r canonical 1-forms which annihilate the pull-back of $\mathcal{E}$; we can write them as $\omega_1,\ldots,\omega_q$, $dt_1,\ldots,dt_r$, since $\mathcal{E}$ is "constant along T".

Thus the dt_j are of no interest to us: what we want to study is the first order variation in T of the forms $\omega_1,\ldots,\omega_q$. So write out each ω_i as a Taylor series in t, with coefficients being forms in x. There will be $(r+1)q$ forms then, $\{\omega_i,\ldots,\omega_q,\omega_j^{t_k}\}$. Similarly on $P_2(\mathcal{Q})$ there will be $(r+1)q^2$ forms $\{\theta_{ij},\theta_{ij}^{t_k}\}$ satisfying $d\omega_i = \sum \theta_{ij} \wedge \omega_j$, $d\omega_i^{t_k} = \sum \theta_{ij}^{t_k} \wedge \omega_j^{t_k}$ and so on.

A moment's thought now shows that the appropriate Lie algebra is L_q augmented as follows: define $\hat{L}_q^{(r)} = L_q \otimes R[t_1\ldots t_r]/(t_it_j)$, $1 \leq i,j \leq r$ with the Lie bracket given by

$$[X_0 + \sum t_iX_i,\ Y_0 + \sum t_iY_i] = [X_0,Y_0] + \sum ([X_0,Y_i] + [X_i,Y_0])t_i.$$

We may regard elements $X_0 + \sum t_iX_i \in \hat{L}_q^{(r)}$ as being the first order part of the Taylor series (along T) of vector fields in L_{q+r}. The bracket then is simply to expand the bracket in L_q, ignoring products in the t_i variables. As a vector space $\hat{L}_q^{(r)} = (r+1)$ copies of L_q, so it has an M-adic topology as well, and hence an M-adic dual $\hat{L}_q^{(r)*}$. This dual space clearly has a basis

$$\{\omega_1,\ldots,\omega_q,t_i\omega_j\},\ \{\theta_{ij},t_k\theta_{ij}\},\ \{\theta_{ijk},t_\ell\theta_{ijk}\}\ \ldots$$

and the differential is extended from that of L_q by requiring that it commute with the t_i.

There is then a natural map $\Phi: \Lambda^*(\hat{L}_q^{(r)*}) \to A^*(P_\infty(\mathcal{Q}))$ which commutes with d and so a map on cohomology. Reasoning as in Chapter III, $H^*(P_\infty(\mathcal{Q})) \simeq H^*(P_1(\mathcal{Q}))$, and so if E has trivial normal bundle, then so does E and we have

$$\hat{\Phi}_E : H^*(\hat{L}_q^{(r)}) \to H^*(M \times T) \xrightarrow{\text{rest}} H^*(M) \ . \quad ^*$$

We call the image of $\hat{\Phi}$ the characteristic ring of the first order variation of E; and we must now show that Im $\hat{\Phi}$ is unchanged if E is replaced by some E' equivalent to it. This follows from the observation that the equivalence class of E remains unchanged if we replace T by any open set containing the base point t_0, and similarly the image of Φ only depends on $M \times U_0$, U_0 being an arbitrarily small open set around t_0: loosely speaking, the class of E and Im Φ only depend on the "germ" of E at $M \times \{t_0\}$ in $M \times T$. For small U_0 around t_0, we can find a bundle map $f: E|_{M\times U_0} \to E'|_{M\times U_0}$ which gives $f: P_1(\mathcal{Q}'|_{M\times U_0}) \to P_1(\mathcal{Q}|_{M\times U_0})$. Under f, the forms $\omega_1,\ldots,\omega_q$ at (x,t_0) and their 1-jets along t will pull back to the corresponding forms on $P_1(E'|_{M\times U_0})$; now the higher P_k are bundles over P_1 with contractible fibres, so choosing trivializations if necessary, we can extend f to each P_k. The pull back of the higher forms will then depend on the homotopy, but in cohomology they will coincide.

Similarly, one has $\Phi: \Lambda^*(L_q^{(r)}, O_{(q)}) \to \Lambda^*(P_\infty(\mathcal{Q})/O_q)$ and $P_\infty(\mathcal{Q})/O_q \sim M \times T$; this follows from the fact that over a suitable open cover of $M \times T$, the transition functions $d\gamma_{\alpha\beta} = \left(\begin{array}{c|c} * & * \\ \hline 0 & I_1 \end{array}\right) \subseteq GL_{q+r}(R)$ and this group has the homotopy type of O_q. Geometrically, this is because E is trivial along T. Thus one has

$$\hat{\Phi}_E : H^*(\hat{L}_q^{(r)}, O(q)) \to H^*(M \times T) \xrightarrow{\text{rest}} H^*(M) \ .$$

---

*So long as T is path-connected, we may restrict to any level $M \times \{t\}$.

and this is again independent of the particular E within the first order equivalence class.

Note that there is a projection $L_q^{(r)} \to L_q$ which is a retraction for the obvious inclusion. In cohomology then one has $H^*(L_q) \to H^*(L_q^{(r)})$, $H^*(L_q,o(q)) \to H^*(L_q^{(r)},o(q))$ and comparing these with $\hat{\Phi}_E$ gives the maps Φ_E of the last section. To define $\frac{\partial}{\partial t} \Phi_E|_{t=0}$, we will define a different map var: $H^*(L_q) \to H^*(L_q^{(r)})$ and one in the relative case. To do this we need to understand $H^*(L_q^{(r)})$ as a module over $H^*(L_q)$ and we do this in the next section.

It is also possible to study variations of integrable subbundles directly in the context of Chapter II. Then one takes the point of view that all the bundles $E_t \subseteq TM$ are isomorphic (over a path connected T) and one fixes an isomorphism $E_y \simeq E_{t_0}$. Thus one can also fix a Riemannian connection. Of course the lie-bracket is varying, and this is reflected in the fact that one puts a varying family of Bott connections on E_{t_0},* and studies the secondary characteristic classes given by each Bott connection. This procedure actually gives a slightly stronger rigidity theorem (i.e., a pointwise one) than what we shall get, but the computations are much more complicated and less conceptual. For this point of view, see Heitsch [12].

---

*A specific example is provided by the example in Chapter I of Bott: the forms θ_A (page 14) are precisely the connection-forms of a varying family of torsion free connections on the trivial bundle spanned by w_A.

2. Variations of a Lie algebra

In this section we shall do the algebra necessary to compute the cohomology of $\hat{L}_q$. We study what one might call the first order "variation" of a real lie algebra $\underline{g}$. This is not quite the same thing as the first order deformation theory and at least heuristically the difference can be described as follows. In studying deformations of $\underline{g}$, one fixes a basis and varies the structure constants; in studying the "variation" one varies the basis and leaves the structure constants fixed* -- this being what is suggested by the lie algebra $\Gamma(E)$ of the previous section. The definitions below will clarify the difference further.

For simplicity, let us begin with the one parameter case. If we wanted to study variation of all orders (in one parameter) we would define a lie bracket on $\underline{g} \otimes R[[t]]$ by

$$(\sum x_i t^i, \sum y_j t^j] = \sum_{k=0}^{\infty} \sum_{i+j=k} [x_i, y_j] \, t^k$$

Notice the differences between this formal power series and the one for deformations. We are however only interested in the first order variation and so we drop all terms (on both sides of the equation) involving t^m, $m \geq 2$. Precisely we define a lie bracket on $\hat{\underline{g}} = \underline{g} \otimes R[t]/t^2$ by

$$[x_0 + x_1 t, \; y_0 + y_1 t] = [x_0, y_0] + ([x_0, y_1] + [x_1, y_0])t \; .$$

---

*Perhaps it is clearer to say that in deformation theory one studies families of (lie) multiplications on a given vector space: here we study a family of vector spaces with the lie multiplication extended naturally from the central fibre.

The functor $\underline{g} \mapsto \hat{\underline{g}}$ is just a change of rings, so the bracket above is clearly a lie bracket. In the same way one can define the n-parameter, first order variation by $\hat{\underline{g}}^{(n)} = \underline{g} \otimes_R R[t_1,\ldots t_n]/(t_i t_j)$ with the bracket given by

$$[x_0 + \sum x_i t_i, y_0 + \sum y_i t_i] = [x_0,y_0] + \sum ([x_0,y_i] + [x_i,y_0])t_i$$

Note that the natural inclusion $\underline{g} \hookrightarrow \hat{\underline{g}}^{(n)}$ given by $x \longmapsto x \otimes 1$ identifies $\underline{g}$ as a subalgebra of $\hat{\underline{g}}^{(n)}$. More importantly, the subspace $\sum_1^n \underline{g}t_i = \underline{g}t$ is an <u>abelian ideal</u> and $(\hat{\underline{g}}^{(n)}/\underline{g}t) = \underline{g}$. Thus there is a split exact sequence of lie algebras

$$0 \to \underline{g}t \to \hat{g}^{(n)} \to \underline{g} \to 0 .$$

In these circumstances, one can compute the cohomology of $\hat{\underline{g}}^{(n)}$ from that of $\underline{g}$ very easily.

Indeed, suppose we have an arbitrary <u>split</u> extension

$$0 \to \underline{a} \to \underline{g} \to \underline{h} \to 0$$

with an abelian kernel $\underline{a}$. Then $\underline{a}$ is an η-module under the splitting, and so $\Lambda^*(\underline{a}^*)$ is a graded η-module. With this understood, one has

<u>Theorem 1</u>. Let $0 \to \underline{a} \to \underline{g} \to \underline{h} \to 0$ be a <u>split</u> extension with abelian kernel $\underline{a}$. Then there are canonical (multiplicative) isomorphisms

$$H^k(\underline{g}) = \sum_{i+j=k} H^i(\underline{h};\Lambda^j(\underline{a}^*)) .$$

In fact the Hochschild-Serre spectral sequence collapses.

Proof. As a vector-space we can write $\underline{g}$ as $\underline{h} \oplus \underline{a}$; because the extension is split we can describe the bracket in $\underline{g}$ in terms of (x,a) and $(x',a') \in \underline{h} \oplus \underline{a}$ as follows:

$$[(x,a),(x',a')] = ([x,x'],x\cdot a'-x'\cdot a)$$

where $x\cdot a$ denotes the action of $\underline{h}$ on $\underline{a}$.

Now again as vector spaces, $\Lambda^k(\underline{g}^*) = \sum_{i+j=k} \Lambda^i(\underline{h}^*) \otimes \Lambda^j(\underline{a}^*)$ and using the bracket rule above we can describe the differential. For notational convenience, set $\Lambda^{i,j} = \Lambda^i(\underline{h}^*) \otimes \Lambda^j(\underline{a}^*) \subseteq \Lambda^{i+j}(\underline{g}^*)$ and let $\hat{d}$ be in the differential in $\Lambda^*(\underline{g}^*)$, d' the differential in $\Lambda^*(\underline{h}^*)$. It is immediate that

(i) $\hat{d}(\Lambda^{1,0}) \subseteq \Lambda^{2,0}$ and in fact $\hat{d}(\xi \otimes 1) = (d'\xi) \otimes 1$

(ii) $\hat{d}(\Lambda^{0,1}) \subseteq \Lambda^{1,1}$ and explicitly $\hat{d}(1 \otimes \alpha)(x,a) = -x\cdot\alpha(a)$

The space $\sum \Lambda^{i,j}$ is of course also the underlying space for the complex which computes $H^*(\underline{h}; \Lambda^*(\underline{a}^*))$. Denoting the differential here by d' again (since on $\sum \Lambda^{i,0}$ it coincides with the earlier definition) we see that $\hat{d} = d'$ on $\Lambda^{1,0}$ <u>and</u> $\Lambda^{0,1}$. Since both differentials are derivations this implies that $\hat{d} = d'$ on all $\Lambda^{i,j}$ and we have proved that the identity map of complexes

$$(\sum \Lambda^{i,j}, d') \to (\Lambda^*(\underline{g}^*), \hat{d})$$

is an isomorphism.

The degeneration of the spectral sequence also follows from the above considerations -- or in fact something <u>slightly</u> weaker. The conditions $\hat{d}(\Lambda^{1,0}) \subseteq \Lambda^{2,0}$ and $\hat{d}(\Lambda^{0,1}) \subseteq \Lambda^{1,1}$

imply that $\hat{d}(\Lambda^{i,j}) \subseteq \Lambda^{i+1,j}$; on the H-S filatration $F^*\Lambda^*(\underline{g}^*)$ relative to the ideal $\underline{a}$ this last containment implies $\hat{d}(F^p\Lambda^k(\underline{g}^*)) \subseteq F^{p+1}\Lambda^{k+1}(\underline{g}^*)$ and thus from the general nonsense of spectral sequences, we have $d_r = 0$ for $r \geq 1$. It remains to identify the E_1-term; this is of course a general fact about extensions and one gets $E_1 = H^*(\underline{h},\Lambda^*(\underline{a}^*))$.

Corollary 1.1. $H^*(\hat{\underline{g}}^{(n)}) = H^*(\underline{g}: \Lambda^*(\underline{g}^* t)) = H^*(\underline{g}: \bigotimes_{i=1}^{n} \Lambda^*(\underline{g}^*))$.

The same considerations apply without essential change to give the following relative version: if $\underline{h}_1 \subseteq \underline{h}$ is a subalgebra, regarded as a subalgebra of $\underline{g}$ by the splitting, then

$$H^*(\underline{g},\underline{h}_1) \simeq H^*(\underline{h},\underline{h}_1: \Lambda^*(\underline{a}^*))$$

and in particular for $\underline{h}_1 \subseteq \underline{g}$

$$H^*(\hat{\underline{g}}^{(n)},\underline{h}_1) \simeq H^*(\underline{g},\underline{h}_1: \bigotimes_{i=1}^{n} \Lambda^*(\underline{g}^*)) .$$

We shall collectively refer to all these assertions as Theorem 1.

In the next section we shall need to know $H^*(L_q:L_q^*)$ explicitly, and so we end this section by describing the result of Gelfand et al. [8].

The cohomology $H^*(L_q:L_q^*)$ will be computed using the Hochschild-Serre spectral sequence for the pair $(L_q,\underline{gl}_q)$ with coefficients in the module L_q^*. The E_1-term is therefore

$$E_1 = H^*(\underline{gl}_q: \Lambda^*((L_q/\underline{gl}_q)^*) \otimes L_q^*)$$

and as $\underline{gl}_q$ is reductive we can apply Theorem 13 of [13] to obtain

$$E_1 = H^*(\underline{gl}_q) \otimes (\Lambda^*(L_q,\underline{gl}_q) \otimes L_q^*)^{\underline{gl}_q}$$

The second factor is the complex for the relative cohomology, and its differential is identified with the differential d_1 in this spectral sequence. Thus the computation of E_2 boils down to computing $H^*(L_q,\underline{gl}_q : L_q^*)$. Now recall that as a $\underline{gl}_q$-module $L_q^* = S^*(R^q) \otimes R^q$: hence in principle the invariant-theory for $\underline{gl}_q$ (which gives the invariants in $S^*(R^q)$ and $\Lambda^*(R^q)$) should compute the second factor in the expression of E_1 above. This is no mean task, but it is fully carried out in [8], and fortunately, the final answer turns out to be simple. $H^*(L_q,\underline{gl}_q : L_q^*)$ is concentrated in degree 2q, and isomorphic* to $H^{2q+1}(L_q)$. Thus

$$E_2^{2q,r} = H^r(\underline{gl}_q) \otimes H^{2q}(L_q,\underline{gl}_q : L_q^*) \ , \qquad r \geq 0$$

and zero otherwise. It follows that the spectral sequence degenerates and

$$H^j(L_q : L_q^*) = H^{2q+1}(L_q) \otimes H^{j-2q}(\underline{gl}_q) \ .$$

Let's work this out for the case $q = 1$. Then using the notation of Chapter III, Section 1, a basis for $\Lambda^*(L_1/\underline{gl}_1)^* \otimes L_1^*$ is given by $\{\alpha_{i_1} \wedge \dots \wedge \alpha_{i_p} \otimes \alpha_k,\ i_j,k \geq -1,\ i_j \neq 0 \text{ for all } j\}$. Recall also that each α_j is an eigenvector for $\underline{gl}_1$ of eigenvalue j. Hence the monomial above is an eigenvector of eigenvalue

---

*This comes about because $H^{2q+1}(L_q)$ is the first nonzero group for L_q. Presumably the map is given by $\Lambda^{2q}(L_q/\underline{gl}_q)^* \otimes L_q^* \to \Lambda^{2q+1}(L_q^*)$ by $\xi \otimes \alpha \to \xi \wedge \alpha$, but this is imbedded in the computations of earlier papers.

$k + \sum_1 i_j$. Thus the invariant subspace is spanned by $\{\alpha_{i_1} \wedge \ldots \wedge \alpha_{i_p} \otimes \alpha_k : k + \sum i_j = 0, i_j \neq 0\}$. Now consider $k + \sum i_j = 0$: if $k = -1$, then $\sum i_j = 1$ and since $i_j \neq 0$ this implies $i_1 = 1$; $i_j = 0$ for $j > 1$. Thus $\alpha_1 \otimes \alpha_{-1}$ is the only invariant with $k < 0$; if $k = 0$, there then $\sum i_j = 0$ so from the earlier computation of $H^*(L_1)$ the invariants $1 \otimes \alpha_0$, $\alpha_{-1} \wedge \alpha_1 \otimes \alpha_0$; if $k > 0$ then $\sum i_j < 0$ and as at most one of the i_j's can be negative, $\alpha_{-1} \otimes \alpha_1$ is the only invariant. To summarize,

$$(\Lambda^*(L_1/\underline{gl}_1)^* \otimes L_1^*)^{\underline{gl}_1} = \left\{\alpha_1 \otimes \alpha_{-1}, \alpha_{-1} \otimes \alpha_1, 1 \otimes \alpha_0, \alpha_{-1} \wedge \alpha \quad \otimes \alpha_0\right\}$$

A quick computation with the differential shows that $\alpha_{-1} \wedge \alpha_1 \otimes \alpha_0 \in H^2(L_q, \underline{gl}_q : L_q^*)$ is the only cochain which contributes to the cohomology, and this gives the desired isomorphism with $H^3(L_q)$ in the obvious way: $\alpha_{-1} \wedge \alpha_1 \otimes \alpha_0 \to \alpha_{-1} \wedge \alpha_1 \wedge \alpha_0$. Thus finally

$$H^j(L_1 : L_1^*) = \begin{cases} R & \text{if} \quad j = 2,3 \\ 0 & \text{for} \quad j = 0,1,4,\ldots \end{cases}$$

There is of course no ring-structure here, but it is evident that as an $H^*(L_1)$-module, it is trivial in positive degrees.

3. Rigidity

We shall define linear maps $\underline{\text{var}}$: $H^p(\underline{g}) \to H^p(\hat{\underline{g}}^{(r)})$ for every $p \geq 1$. These maps essentially measure the derivatives $\partial/\partial t_i$ of a cohomology class evaluated at $t = t_0$. Using this we shall prove the rigidity theorem of Simons and Heitsch that the only classes which can vary continuously are those in the cokernel of the obvious maps $H^*(L_{q+1}, o(q+1)) \to H^*(L_q, o(q))$ and $H^*(L_{q+1}) \to H^*(L_q)$ induced by inclusion.

The general case of r-parameters follows from the 1-parameter case mutatis mutandi, and so we mainly discuss the 1-parameter case. For each $p \geq 0$ we have a map of $ad(\underline{g})$-modules, $w: \Lambda^p(\underline{g}) \otimes \underline{g} \to \Lambda^{p+1}(\underline{g})$ given by wedge product: $w(x \otimes y) = x \wedge y$, where $x \in \Lambda^p(\underline{g})$, $y \in \underline{g}$. Its transpose is then also a map of $ad(\underline{g})$ modules which we can take to be $v: \Lambda^{p+1}(\underline{g}^*) \to \Lambda^p(\underline{g}^*) \otimes g^*$. (Recall that in case $\underline{g} = L_q$, g^* is not the full linear dual; however because L_q^* is the direct limit of finite-dimensional spaces, one can commute * and exterior powers.) The map v can be specified thus: for $\phi \in \Lambda^{p+1}(\underline{g}^*)$, $x \in \Lambda^*(\underline{g})$, $y \in \underline{g}$,

$$v(\phi)(x \otimes y) = \phi(x \wedge y)$$

or explicitly, for $\xi_{i_0}, \ldots, \xi_{i_p} \in \underline{g}^*$,

$$v(\xi_{i_0} \wedge \ldots \wedge \xi_{i_p}) = \sum_{j=0}^{p} (-1)^{p-j} \xi_{i_0} \wedge \ldots \hat{\xi}_{i_j} \ldots \wedge \xi_{i_p} \otimes \xi_{i_j}$$

Now we have differentials in the complexes $\Lambda^*(\underline{g}^*)$, $\Lambda^*(\underline{g}^*) \otimes \underline{g}^*$, and we claim that v commutes with them.

Proposition 1. For each $p \geq 0$, $v: \Lambda^{p+1}(\underline{g}^*) \to \Lambda^p(\underline{g}^*) \otimes g^*$ commutes with the differentials and so induces a map $v: H^{p+1}(\underline{g}) \to H^p(\underline{g}:\underline{g}^*)$.

Proof. For $p = 0$, v is the identity map, so clearly commutes with d. We now argue inductively, using the fact that in lie algebra cohomology (with any coefficients) the differential can be defined inductively in terms of the $\underline{g}$-action and interior products. Now v is a map of $\underline{g}$-modules by construction, so it suffices to show that for every $x \in \underline{g}$, $i_x v = v i_x$. This can be checked directly from the definition, or more computationally, from the explicit formula above. □

If $\underline{h} \subseteq \underline{g}$ is a subalgebra, there is a relative version of v which induces maps $v: H^{p+1}(\underline{g},\underline{h}) \to H^p(\underline{g},\underline{h}:\underline{g}^*)$. As for the multiplicative properties of v, it satisfies the derivation rule

$$v(\alpha \wedge \beta) = \alpha \wedge v(\beta) + v(\alpha) \wedge \beta$$

where the wedge product in the second term has to be interpreted as follows: if $v(\alpha) = \alpha_1 \otimes \alpha_2$, $v(\alpha) \wedge \beta = (-1)^b(\alpha_1 \wedge \beta)\otimes \alpha_2$, $b = \deg \beta$.

<u>Definition</u>. The variation homomorphism is defined as the composite $\underline{\text{var}}: H^{p+1}(\underline{g},\underline{h}) \xrightarrow{v} H^p(\underline{g},\underline{h}:\underline{g}^*) \hookrightarrow H^{p+1}(\hat{\underline{g}},\underline{h})$ where the second inclusion is the one given in the previous section.

Since the inclusion $H^p(\underline{g},\underline{h}:\underline{g}^*) \hookrightarrow H^{p+1}(\hat{\underline{g}},\underline{h})$ is given by the obvious identification of complexes, one can give an explicit formula for $\underline{\text{var}}$ at the cochain level. Taking $\underline{h} = (0)$ for simplicity, we get

$$\underline{\text{var}}\,(\xi_{i_0} \wedge \ldots \wedge \xi_{i_p}) = \sum (-1)^{p-j}\, \xi_{i_0} \wedge \ldots \hat{\xi}_{i_j} \wedge \ldots \wedge \xi_{i_p} \otimes \xi_{i_j} t$$

To conclude this formal algebraic set-up we show how it extends to the n-parameter case. First, $H^p(\underline{g}: \Lambda^1(\underline{g}^* t)) = \bigotimes_1^n H^p(\underline{g}:g^* t_i)$, and for each summand in this group we have a map $v_i: H^{p+1}(\underline{g})$

$\to H^p(\underline{g}:\underline{g}^* t_i)$. Thus finally we have the variation map

$$\text{var}: H^{p+1}(\underline{g}) \xrightarrow{\sum v_i} \oplus H^p(\underline{g}: \Lambda^1(\underline{g}^* t)) \hookrightarrow H^{p+1}(\hat{\underline{g}}^{(n)}) .$$

We now return to the geometry ($\underline{g} = L_q$) and for simplicity we only discuss the one parameter case in detail. So let $E \subseteq T\,M$ be an integrable subbundle of codimension q, and $\mathcal{E}$ a 1-parameter variation of E. Then we have the Gelfand-Fuks maps

$$\phi_{E_t} : H^*(L_q, o(q)) \to H^*(M)$$

where $E_t = \mathcal{E}|_{M\times\{t\}}$ and

$$\hat{\phi}_{\mathcal{E}} : H^*(\hat{L}_q, o(q)) \to H^*(M)$$

The first assertion about the variation homomorphism is then

<u>Proposition 2</u>. For every cohomology class $\alpha \in H^p(L_q, o(q))$ we have

$$\hat{\Phi}_{\mathcal{E}}(\text{var}(\alpha)) = \frac{\partial}{\partial t} (\Phi_{E_t}(\alpha)) \Big|_{t=0}$$

Similarly for foliations with trivial normal bundle.

<u>Proof</u>. There is really nothing to prove once the definitions of these maps have been recalled -- and then this identity is seen to hold at the chain level. Since $\partial/\partial t$ commutes with the differentials in the cochain complexes, the identity holds in cohomology as well.

At the chain level, the maps Φ_{E_t} and $\hat{\Phi}_{\mathcal{E}}$ are defined on the first exterior power of the duals of the appropriate lie algebras, and then extended multiplicatively; moreover $\frac{\partial}{\partial t}(\ )\Big|_{t=0}$ and var both satisfy Leibniz rules.† Hence it suffices to check the identity on elements of L_q^*. Now $\mathrm{var}: L_q^* \xrightarrow{v} \Lambda^0(L_q^*) \otimes L_q^* \hookrightarrow L_q^*$ sends an element $\gamma \in L_q^*$ to $\gamma t \in \hat{L}_q^*$. By construction, $\hat{\Phi}$ sends the elements $\omega_i t, \theta_{ij} t, \ldots$ to the 1-jet along t at 0 of the forms defining E; this is precisely the operation $\frac{\partial}{\partial t}(\ )\Big|_{t=0}$, and the result follows.□

We will now describe the kernel of var; in view of Proposition 2 above, this kernel is the set of cohomology classes which are "universally" or *a priori* rigid. To state the result, denote the natural inclusion $L_q \to L_{q+1}$ by i_q.

Proposition 3. The following sequences are exact.

$$H^*(L_{q+1}) \xrightarrow{i_q^*} H^*(L_q) \xrightarrow{\mathrm{var}} H^*(L_q)$$

and

$$H^*(L_{q+1}, O(q+1)) \xrightarrow{i_q^*} H^*(L_q, O(q)) \xrightarrow{\mathrm{var}} H^*(\hat{L}_q, O(q)) .$$

Proof. The proofs in the two cases are parallel, so we only discuss the first. The exactness depends upon the explicit calculations of the cohomology groups and the maps, and we shall describe it briefly: but one half of the exactness, namely that $\mathrm{var} \circ i_q^* = 0$, has a simple geometric explanation which we now give.

So suppose $E \subseteq TM$ is an integrable subbundle of codimension q,‡ and E is a 1-parameter variation of E. Then on $M \times R$ we

† See above.

‡ With trivial normal bundle, since we are discussing the first sequence.

can construct the "constant" foliations $E'_t = E_t \times \{0\} \subseteq TM \times TR$ which are of codimension q+1. As integrable subbundles of $T(M \times R)$ all the E'_t are integrably homotopic. Indeed, if $\pi\colon M \times R \times R \to M \times R$ is the projection given by $\pi(m,s,t) = (m,t)$ then $\pi^*(E)$ is an integrable subbundle of codimension q+1 which restricts at $M \times R \times \{t\}$ to E'_t. Thus the Gelfand-Fuks maps

$$\Phi_{E'_t}\colon H^*(L_{q+1}) \longrightarrow H^*(M \times R)$$

are all independent of t. On the other hand, if $i_t\colon M \to M \times R$ is the map $i_t(m) = (m,t)$ then by construction the following diagram commutes:

$$\begin{array}{ccc} H^*(L_{q+1}) & \xrightarrow{i_q^*} & H^*(L_q) \\ \Phi_{E'_t}\big\downarrow & & \big\downarrow\Phi_{E_t} \\ H^*(M\times R) & \xrightarrow{i_t^*} & H^*(M) \end{array}$$

and thus the composite $\Phi_{E_t} \circ i_q^*$ is independent of t. Applying Proposition 2, we see that

$$\hat{\Phi}(\text{var} \circ i_q^*(\beta)) = \frac{\partial}{\partial t} (\Phi_{E_t} i_q^*(\beta))\Big|_{t=0} = 0$$

for any $\beta \in H^*(L_{q+1})$. This shows that in any geometric case at least $\text{var} \circ i_q^* = 0$.

Now to the algebraic argument. Since <u>var</u> is the composite of v and the inclusion $H^*(L_q : L_q^*) \to H^*(\hat{L}_q)$ we may as well show $v \circ i_q^* = 0$. Although we did not explicitly state it when discussing v, it is clear that v is functorial on the category of lie algebras and also that it can be defined with coefficients -- i.e., there are maps $v_M\colon H^*(\underline{g}:M) \to H^*(\underline{g}\colon \underline{g}^* \otimes M)$ for any $\underline{g}$

and $\underline{g}$-module M. These two facts show that the map $v: H^*(L_q) \to H^*(L_q : L_q^*)$ can actually be defined on the spectral sequences for these groups: thus, there is a map on the E_1-terms

$$v_1: H^*(\underline{gl}_q: \Lambda^*(L_q, \underline{gl}_q)) \to H^*(\underline{gl}_q: \Lambda^*(L_q, \underline{gl}_q : L_q^*))$$

which gives v on the E_∞-term. This v_1 can be calculated using the elements u_i, c_i in the domain which give a basis for $H^*(L_q)$ and the fact that the E_2-term of the range is concentrated in fiber degree 2q. The paper of Gelfand et al. [8] gives actual cocycles representing this group and from this calculates $v_2(u_I c_J)$ as an explicit cocycle. Now the image i_q^* is similarly calculated at the E_1-term in terms of the u_i, c_i again and a comparison of these computations yields the result.

Appendix I. Connections, curvature and characteristic classes

We shall set down the basic facts concerning the Chern-Weil theory of characteristic classes as it is used in the body of these notes. This appendix is simply intended to guide the reader, and help him locate the relevant portions of the theory in the literature: see Milnor [17] for the first three parts, and Kobayashi-Nomizu [15] and H. Cartan [5] for the last section. However, for the reader who likes to relish his mathematics, there is no substitute for reading Chern's original papers [6a,b].

1. Connections

M is a smooth, paracompact manifold, and $V \to M$ a smooth real vector bundle over M. (The theory for smooth complex bundles can be worked out from the real case by a little linear algebra.) $\Gamma(V)$ denotes the space of smooth sections of V.

A connection on ∇ is a real linear map

$$\nabla : \Gamma(TM) \times \Gamma(V) \to \Gamma(V)$$

satisfying the conditions

(i) $\nabla_{fX}(s) = f\nabla_X(s)$, $f \in C^\infty(M)$

(ii) $\nabla_X(gs) = g\nabla_X(s) + X(g)\cdot s$, $g \in C^\infty(M)$

These conditions say that ∇ is a "local" operator, and in fact a first order differential operator: thus if X or s vanishes near a point, so does $\nabla_X(s)$. Given this, one could equally well define ∇ as an additive map of sheaves (of sections of TM and V) satisfying (i) and (ii).

Connections always exist in the differentiable setting: if $U \subseteq M$ is an open set over which V is trivial, we may choose local trivializing sections $s_1,\ldots,s_k$ and define $\nabla_X(s_i) = 0$ for every $X \in \Gamma(TM|_U)$, extended by rule (ii) to all $s = \sum g_i s_i$, so that $\nabla_X(s) = \sum_1^k X(g_i)s_i$. Choosing a partition of unity subordinate to a covering of M by open sets trivializing V, ∇ extends to global sections.

Sometimes it is more convenient to adopt a "dual" definition of a connection, as a real linear map

$$D: \Gamma(V) \longrightarrow \Gamma(TM^* \otimes V)$$

satisfying

$$D(fs) \quad = \quad f\, D(s) + df \otimes s \quad \text{for} \quad f \in C^\infty(M)$$

This is entirely equivalent to the previous definition by the identity (read either way)

$$D(s)(X) = \nabla_X(s) \qquad \text{for} \qquad X \in \Gamma(TM)\ .$$

In terms of the first definition, one can define the "local connection form" of a connection <u>with respect to a basis</u>, as follows: let $U \subseteq M$ for an open set, and $s_1,\ldots,s_k \in \Gamma(V|_U)$ sections which form a basis for the fiber over every point: then $\nabla_X(s_i) = \sum_{j=1}^k \eta_{ij}(X)s_j$. Thus the $(k\times k)$-matrix (η_{ij}) may be regarded as a 1-form with values in the $(k\times k)$-matrices: that is, we can regard ∇ (by the exponential law) as a map

$\nabla: \Gamma(TM) \to \mathrm{Hom}(\Gamma(V), \Gamma(V))$, and (η_{ij}) is the local expression of this map.

Note however, that (η_{ij}) does not transform like a section of the bundle $\mathrm{Hom}(TM, \mathrm{End}_R(V))$ because ∇ is not linear over functions: hence in this setting there is no global connection form on M. For example, if $f_1, \ldots, f_k \in C^\infty(M)$ are functions not zero over U, and one uses the basis $\sigma_i = f_i s_i$ instead, then

$$\nabla_X(\sigma_i) = f_i \nabla_X(s_i) + X(f_i)s_i = \sum (f_i \eta_{ij}(X) s_j) + df_i(X) s_i$$
$$= f_i\left(\sum \eta_{ij}(X) s_j + \frac{df_i(X)}{f_i} s_i\right) .$$

A similar transformation rule applies if the basis $\{s_i\}$ is transformed by a matrix of functions A, invertible over U.

If the bundle V has a Riemannian metric $< \, , >$ then a connection ∇ is said to be Riemannian (with respect to $< \, , >$) if for all $X \in \Gamma(TM)$ and $s, s' \in \Gamma(V)$ it satisfies

$$<\nabla_X(s), s'> + <s, \nabla_X(s')> = X(<s,s'>)$$

In general there is no unique Riemannian connection associated to a metric, but one always exists. Over an open set $U \subseteq M$ on which V is trivial, choose an orthonormal basis $s_1, \ldots, s_n$ and put $\nabla_X(s_i) = 0$; then patch together by a partition of unity.

2. Curvature

Given a connection on a vector bundle $V \to M$, its curvature is a bundle map $k_\nabla: \Lambda^2(TM) \to \mathrm{Hom}(V,V)$ constructed as follows.

Let $\nabla: \Gamma(TM) \times \Gamma(V) \to \Gamma(V)$ be the connection: as remarked above, we may regard it as a map $\nabla: \Gamma(TM) \to \Gamma(\mathrm{Hom}(V,V))$. Now both the domain and range are lie algebras: the domain under the usual lie bracket of vector fields, the range under the usual $[A,B] = AB - BA$. One can therefore ask whether ∇ is a lie algebra homomorphism; or more precisely one can measure its deviation from being one by setting

$$\kappa: \Gamma(TM) \times \Gamma(TM) \to \Gamma(\mathrm{Hom}(V,V))$$

$$\kappa(X,Y) = \nabla_X\nabla_Y - \nabla_Y\nabla_X - \nabla_{[X,Y]}$$

Notice first of all that κ is skew-symmetric in X,Y -- i.e., $\kappa(X,Y) = -\kappa(Y,X)$ -- and second, that κ is linear over functions in all arguments. Thus in the third argument, for example

$$\begin{aligned}\kappa(X,Y)(fs) &= \nabla_X\nabla_Y(fs) - \nabla_Y\nabla_X(fs) - \nabla_{[X,Y]}(fs)\\ &= f\nabla_X\nabla_Y(s) + X\,Y(f)s - f\nabla_Y\nabla_X(s) - YX(f)s\\ &\quad - f\nabla_{[X,Y]}(s) - [X,Y](f)s\\ &= f\,\kappa(X,Y)(s)\end{aligned}$$

Hence κ defines a bundle map $k_\nabla: \Lambda^2(TM) \to \mathrm{End}_R(V)$.

Dually, if we start with the definition of a connection as a map $D: \Gamma(V) \to \Gamma(TM^* \otimes V)$, then one can define the "prolongations" of D by setting $D_{(p)}: \Gamma(\Lambda^p(TM^*) \otimes V) \to \Gamma(\Lambda^{p+1}(TM^*) \otimes V)$ to be

$$D_{(p)}(\omega \otimes s) = d\omega \otimes s + (-1)^p\omega \wedge D(s)$$

In these terms, the curvature measures the obstruction to the above sequence being a complex: that is, it appears as a map $\kappa: \Gamma(V) \to \Gamma(\Lambda^2(TM^*) \otimes V)$ such that $D_{(1)} \circ D_{(0)}(s) = \kappa(s)$ and then, by algebraic manipulation, $D_{(p)} \circ D_{(p-1)}(\omega \otimes s) = \omega \wedge \kappa(s)$, $p \geq 2$.

As above κ is linear over functions, so defines a bundle map $k_D: V \to \Lambda^2(TM^*) \otimes V$, which may be seen to be equivalent to the earlier k_∇.

The reader is advised not to look too hard for a geometric picture which identifies these abstractly defined tensors with the intuitive notion of curvature. The intuitive notion really comes into its own only when k is the curvature of the Levi-Civita connection for the induced metric on a manifold imbedded in a Euclidean space.

When the bundle V carries a Riemannian metric $< , >$ and ∇ is a Riemannian connection associated to $< , >$, then the curvature $k_\nabla(X,Y)$ satisfies

$$<k_\nabla(X,Y)s,s'> + <s,k(X,Y)s'> = 0$$

-- i.e., k (X,Y) is a skew-symmetric matrix for each pair (X,Y).

Given an open set $U \subseteq M$ over which V is trivial, and $\eta = (\eta_{ij})$ a local connection form with respect to <u>any</u> basis for $\Gamma(V|_U)$ one has the basic identity $k_\nabla = d\eta - \eta \wedge \eta$. This is the so-called "first structural equation" and is very useful in computations. It is easily verified thus: Let $s_1, \ldots, s_n$ be a basis of V over $U \subseteq M$. The connection form (η_{ij}) is then given by $D(s_i) = \sum_j \eta_{ij} \otimes s_j$. The curvature

$$k_D(s_i) = D_1 \circ D_0(s_i) = D_1(\sum \eta_{ij} \otimes s_j)$$

$$= \sum d\eta_{ij} \otimes s_j - \sum \eta_{ij} \wedge D(s_j)$$

$$= \sum_j d\eta_{ij} \otimes s_j - \sum_{j,k} (\eta_{ij} \wedge \eta_{jk}) \otimes s_k$$

so

$$k_D = d(\eta_{ij}) - (\eta_{ij}) \wedge (\eta_{jk}) .$$

We shall see later on that in the context of principal bundles, the connection form can be globally defined, and then this equation has a global version.

3. Characteristic classes

Fix a bundle $V \to M$ with a connection ∇, and let $k_\nabla : \Lambda^2(TM) \to \mathrm{End}_R(V)$ be its curvature. Locally, over a trivializing neighborhood for V, k_∇ is a "matrix-valued 2-form". To get ordinary forms out of it, locally one may apply any C^∞ function to the matrix k_∇ over U: however, to get a global form, one can only apply functions which are invariant under change of basis. There is no loss of generality in restricting oneself to invariant polynomials, and so if $\phi \in I(\underline{gl}_k)$ is a homogeneous invariant polynomial of degree d on the $(k\times k)$-matrix space, $\phi(k_\nabla)$ is a 2d-form on M.

The basic theorems of the Chern-Weil theory are then the following two.

Theorem 1. $\phi(k_\nabla)$ is a closed form, for every $\phi \in I(\underline{gl}_k)$.

Theorem 2. If ∇_1 and ∇_2 are two connections on V with curvatures k_1, k_2 then for any $\phi \in I(\underline{gl}_k)$, $\phi(k_1) - \phi(k_2)$ is exact.

For the proof of the first theorem, we refer to Milnor [17]: the second one is needed in Chapter II, and is proved there. It remains only to describe what the invariant polynomials on $\underline{gl}_k$ are.

Let us begin with some simple examples. If A is a k×k-matrix, then clearly trace(A), and more generally trace (A^p) for $p = 1,2,3,...,$ are invariant polynomials; if A is in upper-triangular form, then trace(A^p) is simply the sum of the pth powers of the diagonal entries (= eigenvalues). Another set is provided by the following formula: let t be an indeterminate, and consider det $(tI + A)$ as a polynomial in t. It can be expanded as det $(tI + A) = \sum_{i=0}^{k} c_i(A)t^{k-i}$, and clearly the $c_i(A)$ are invariant polynomials; in fact $c_1(A) = \text{trace}(A)$, $c_k(A) = \det A$ and more generally, $c_i(A) = i^{th}$ elementary symmetric function of the eigenvalues. It turns out that either of the sets $\{\text{trace}(A),...,\text{trace}(A^k)\}$ or $\{c_1(A),...,c_k(A)\}$ generate the set of invariant polynomials, and they are algebraically independent. That is, we have

Proposition. $I(\underline{gl}_k)$ is isomorphic as a ring to $R[x_1,...,x_k]$, the polynomial ring in k variables, where the c_i may be chosen to represent the x_i.

For a nice proof, see the appendix to Bott's Mexico lectures [2].

The Chern-Weil theory may be regarded as giving a ring homomorphism

$$\text{C-W: } I(\underline{gl}_k) \to \Gamma(\Lambda^*(TM^*)) \; \dagger$$

actually mapping into the forms of even degree. This map depends on the choice of connection; but by Theorem 1 we may pass to the cohomology, and by Theorem 2, the image in cohomology is independent of the connection. We can make a further improvement by choosing the curvature of a Riemannian connection: then the curvature is skew symmetric, so the odd c_i's vanish. Hence the image of $I(\underline{gl}_k)$ in cohomology is non-zero only in degrees 4i; the images of the c_i (up to factors of π and other universal constants) are the Pontryagin classes.

4. H. Cartan's formalism

For purposes of computing characteristic classes on lie groups, homogeneous spaces and so on, it is useful to have a quite different formulation of connections and cur ature from that developed above. We give a breezy discussion of this new formulation, relating it to the earlier definitions.

Fix a vector bundle $V \to M$, and assume M is connected. A connection ∇ on V gives a way of "comparing" the fibers of V at two different points, as follows. Let $x_0, x_1 \in M$ be given, $\sigma\colon [0,1] \to M$ a smooth curve from x_0 to x_1 and X = velocity vector-field of σ. We say that $v_0 \in V_{x_0}$ and $v_1 \in V_{x_1}$ are

†Bott has recently given a construction of this map which does not bring in connections at all: see "On the Chern-Weil Homomorphism and the Continuous Cohomology of Lie Groups," in Adv. in Math., Vol. 11 (1973), pp. 289-303.

parallel along σ if there is a section $s: \sigma([0,1]) \to V$ (equivalently a lifting of σ) satisfying

(i) $s(\sigma(0)) = s(x_0) = v_0$, $s(\sigma(1)) = s(x_1) = v$,

(ii) $\nabla_X(s) = 0$

This gives a linear isomorphism of V_{x_0} with V_{x_1} -- of course, depending on σ. Explicitly, to each $\sigma: [0,1] \to M$, going from x_0 to x_1 , there is an isomorphism $\phi_\sigma: V_{x_0} \to V_{x_1}$ defined as follows: $\phi_\sigma(v_0) = s(1)$ where $s: \sigma([0,1]) \to M$ is the unique section† parallel along σ, with initial point v_0. This assignment is linear by (ii) and it is an isomorphism because $\phi_{\sigma^{-1}} = (\phi_\sigma)^{-1}$, where σ^{-1} is the "same" path running backwards -- i.e., $\sigma^{-1}(t) = \sigma(1 - t)$ for $t \in [0,1]$. Conversely if we have a notion of parallelism along all smooth curves, we can recover a connection on V as follows. Given a vector field $X \in \Gamma(TM)$ and a global section $s \in \Gamma(V)$ we define $\nabla_X(s)$ at a point m by the formula

$$\nabla_X(s) = \lim_{t \to 0} \frac{\phi_{-t}(s(\xi_t(m)) - s(m)}{t}$$

where ξ_t is the local flow induced by X starting at m, and ϕ_{-t} is the parallel transport along the curve ξ_{-t}. Thus $\nabla_X(s)$ is, so to speak, "the infinitesimal parallel displacement" of s along X.

All this can be put more algebraically in terms of principal bundles. We shall only discuss the case of GL-bundles, and leave the reader to sort out the details where there is a reduc-

†Uniqueness follows from uniqueness for ODE with given initial values.

tion of the structural group. So let $\pi: P(V) \to M$ be the principal GL_k-bundle of V. On P(V) we have an exact sequence of vector bundles

$$0 \longrightarrow F \longrightarrow T(P(V)) \xrightarrow{d\pi} \pi^*(TM) \longrightarrow 0$$

where F is the bundle of tangents along the fibers.
A connection in V is equivalent to a "GL_k-invariant" splitting of this sequence as we now show.

Suppose we have a notion of parallel transport in V; then for every vector field X on M we want to define its "horizontal lift" $\tilde{X}$ on P(V) so that $d\pi(X) = \tilde{X}$, and so that $\tilde{X}$ will be a GL_k-invariant along the fibers. So fix a point $(a,m) \in P(V)$ where a is a frame for V_m. Let ξ_t denote the local flow on M induced by X and starting at m. Along ξ_t we can parallel-transport the frame a to get frames a_t for $V_{\xi_t(m)}$. This gives a smooth curve $\tilde{\xi}_t = (a_t, \xi_t(m))$ in P(V) starting at (a,m) and we define $\tilde{X}$ to be the velocity vector of the curve $\tilde{\xi}_t$ at (a,m). If (a',m) is another point lying over m, there is a unique $g \in GL_k$ such that $a' = g \cdot a$. Since parallel transport is a linear isomorphism of fibers, $a_t' = g \cdot a_t$ and so we see that $dg(\tilde{X}_a) = \tilde{X}_{ga}$. Thus parallel transport in V defines a linear map $\sigma: \pi^*(TM) \to T(P(V))$, $\sigma(X) = \tilde{X}$ such that

(i) $d\pi \circ \sigma(X) = X$; (ii) $\sigma(X_{ga}) = dg\, \sigma(X_a)$.

Conversely given a GL_k-invariant splitting σ as above, we can define parallel transport in V. Let $c: [0,1] \to M$ be a smooth curve, a_0 a frame for $V_{c(0)}$. We can lift the velocity field dc/dt to $\sigma(dc/dt)$, and thus define its integral curve

γ: $[0,1] \to P(V)$ starting at (a_0,m). Clearly $\pi \circ \gamma = c$, and we define the parallel transport along γ to be a_t where $\gamma(t) = (a_t, c(t))$. This is clearly a linear isomorphism of the fibers $V_{c(t)}$ satisfying all the required conditions.

Now the splitting $\sigma: \pi^*(TM) \to TP(V)$ defines the horizontal and vertical components of an arbitrary vector field $Y \in \Gamma(T(P(V)))$: $Y_h = \sigma \circ \pi(Y)$, $Y_v = Y - Y_h$. In general, the bracket of horizontal vector fields will not be horizontal: or more formally, $\Gamma(\sigma)$: $\Gamma(\pi^*(TM)) \to \Gamma(T(P(V)))$ will not be a lie algebra homomorphism. This corresponds precisely to the fact that the connection defined in Section 1 was not a lie algebra homomorphism in general. In any case, if we take $X_1, X_2 \in \Gamma(\pi^*(TM))$ the vertical component $[\sigma(X_1), \sigma(X_2)]_v$ gives a map κ_σ: $TM \times TM \to \underline{gl}_k$, as follows. By definition, $[\sigma(X_1), \sigma(X_2)]_v$ is a section of F: from the GL_k-invariance of σ we see that we can identify it with a GL_k-invariant field along the fiber, and thus with an element of $\underline{gl}_k$.† It is not hard to show that $\kappa_\sigma(X_1, X_2) = [\sigma(X_1), \sigma(X_2)]_v$ is precisely the curvature introduced earlier.

Cartan begins by dualizing the whole set-up. Thus instead of a splitting σ as above, he starts out with the dual splitting τ: $TO(V) \to F$ (i.e., $\tau = 1 - \sigma \circ \pi$) which by GL_k-invariance may be regarded as a map τ: $TP(V) \to \underline{gl}_k$. This is the global version of the connection form defined earlier. Taking linear

†The fiber over a point is of course only a principal homogeneous space of GL_k -- the space of invariant vector fields is however well defined, and isomorphic to the lie algebra of the group.

duals we get a map on sections

$$\omega\colon \underline{gl}_k^* \to A^1(P(V))$$

and denoting by θ_{ij} the obvious basis of $\underline{gl}_k^*$, $\omega_{ij} = \omega(\theta_{ij})$ are the components of the connection form. ω extends to a graded algebra homomorphism

$$\omega\colon \Lambda^*(\underline{gl}_k^*) \to A^*(P(V))$$

and the fact that ω comes from a GL_k-invariant splitting can be expressed as follows: Consider $A \in \underline{gl}_k$ regarded as a vector field on $P(V)$ along the fibers: then

(i) $i_A\omega(x) = \omega(i_A x)$ and (ii) $ad_A\omega(x) = \omega(ad_A x)$,

where the first condition expresses the fact that ω comes from a splitting and the second expresses the $\underline{gl}_k$-invariance.

Despite all these restrictions on ω, it may still not commute with the differentials. Indeed, it is again the curvature κ_σ which accounts for this discrepancy. To see this, let $\{\theta_{ij}\} \subseteq \underline{gl}_k$ be the usual basis and recall that

$d\theta_{ij} = \sum_{\ell=1}^{k} \theta_{i\ell} \wedge \theta_{\ell j}$: writing $\omega_{ij} = \omega(\theta_{ij})$ we get

$$d\omega(\theta_{ij}) - \omega(d\theta_{ij}) = d\omega_{ij} - \sum \omega_{i\ell} \wedge \omega_{\ell j}$$

and by the first structural equation the last expression is precisely κ_{ij}. We now combine all this information into a "super homomorphism" as follows. For $x \in \underline{gl}_k^*$, put $\kappa(x) = d\omega(x) - \omega(dx) \in A^2(P(V))$ and extend this to a graded algebra homomorphism $\kappa\colon S^*(\underline{gl}_k^*) \to A(P(V))$ where $S^j(*)$ is assigned degree 2j. Then if we set $W(\underline{gl}_k) = \Lambda^*(\underline{gl}_k^*) \otimes S^*(\underline{gl}_k^*)$

we can define

$$\psi: W(\underline{gl}_k) \to A^*(P(V)) \quad \text{by} \quad \psi(x \otimes y) = \omega(x) \wedge \kappa(y) .$$

For $A \in \underline{gl}_k$ one can introduce the operations i_A, ad_A in the Weil algebra $W(\underline{gl}_k)$; indeed both the exterior and symmetric algebras are $\underline{gl}_k$-modules, hence so is the tensor product, and this gives the action of ad_A. One requires i_A to be a derivation acting on $\Lambda^*(\underline{gl}_k^*)$ as usual, and annihilating $1 \otimes S^*(\underline{gl}_k^*)$. ψ is then a map which commutes with i_A, ad_A.

In any (graded) algebra which is a module over $\underline{g}$ with interior product, one can define a differential δ satisfying $i_A\delta = ad_A - \delta i_A$.[†] Thus we get a differential in $W(\underline{gl}_k)$ and because ψ commutes with i_A, ad_A it satisfies $\psi(\delta(x \otimes y)) = d(\psi(x \otimes y))$.

We can now define the Chern-Weil map. The differential forms on M, $A^*(M)$ can be identified with the subalgebra of differential forms $A^*(P(V))$ which are $\underline{gl}_k$-basic: that is, $\underline{gl}_k$-invariant, and annihilated by i_A. The corresponding basic elements in $W(\underline{gl}_k)$ are simply the invariant symmetric elements: $S^*(\underline{gl}_k^*)^{\underline{gl}_k}$. Now the basic elements in $W(\underline{gl}_k)$ are cocycles for δ as one sees from the defining equation: hence so are their images in $A^*(P(V))$ and we have

$$\bar{\psi}: S^*(\underline{gl}_k^*)^{\underline{gl}_k} \to \pi^*(A^*(M)) \subseteq A^*(P(C))$$

takes its image into closed forms. This is then the Chern-Weil homomorphism. It is an algebraic calculation to show that in

[†]See Appendix II.

cohomology the image of $\bar{\psi}$ is independent of the connection form ω used to construct it.

One final remark concerning homogeneous spaces. Consider a lie group G, with a closed subgroup $H \subseteq G$, and consider the fiber-bundle $H \to G \xrightarrow{\pi} G/H$. Suppose that the cotangent bundle (G/H) admits a G-invariant connection. Then we can regard the above fiber bundle as the principal bundle of T(G/H) with a connection and the Weil map ψ then goes directly into the <u>invariant forms</u> on $\underline{g}$. The assertion is trivial to check, essentially by definition of invariant connection.

Appendix II. Cohomology of Lie algebras

1. Consider a connected lie group G. In computing its deRham cohomology from the algebra $A^*(G)$ of all smooth forms, it is natural to ask what the left invariant forms contribute, and this leads to an abstract (algebraic) definition of the cohomology of any lie algebra, as we shall now show. So let $\underline{g}$ be the lie algebra of G; then the dual vector space $g^* = \mathrm{Hom}(\underline{g},R)$ can be identified with the space of left-invariant one-forms, and the exterior algebra $\Lambda_R^*(\underline{g}^*)$ with the subalgebra $A^*(G)$ of all left-invariant forms. In fact $\Lambda_R^*(\underline{g}^*)$ is a subcomplex of $A^*(G)$; for if ω is a left invariant form, and $L_g: G \to G$ is left multiplication by g, then by the functoriality of $d: A^p(G) \to A^{p+1}(G)$ we see that

$$L_g^*(d\omega) = d(L_g^*\omega) = d\omega$$

so $d\omega$ is left invariant as well.

Now let us see how we can describe d on the subcomplex $\Lambda_R^*(\underline{g}^*)$. It suffices to describe it on $\Lambda^0(\underline{g}^*)$ and $\Lambda^1(\underline{g}^*)$, for d is a derivation. $\Lambda^0(\underline{g}^*)$ consists of the constant functions, so d is zero. If $\omega \in \Lambda^1(\underline{g}^*)$ it suffices to evaluate $d\omega$ on a pair $X,Y \in \underline{g}$, since we know that $d\omega \in \Lambda^2(\underline{g}^*)$:

$$d\omega(X,Y) = X\omega(Y) - Y\omega(X) - \omega([X,Y])$$

and the first two terms drop out since $\omega(X)$, $\omega(Y)$ are constant functions. Thus d (X,Y) = - ([X,Y]) and so on an arbitrary p-form $\phi \in \Lambda^p(\underline{g}^*)$, $d\phi(X_0,\ldots,X_p) = \sum_{i<j} (-1)^{i+j} \phi([X_i,X_j], X_0,\ldots,\hat{X}_i,\ldots,\hat{X}_j,\ldots,X_p)$.

This last formula makes sense for any lie algebra L defined over any ring A, and so we can define its cohomology $H^*(L)$ as follows.

<u>Definition 1</u>. Let L be a lie algebra over the ring A. Its cohomology, by definition is the cohomology of the complex $(\Lambda_A^*(L^*),d)$ where $L^* = \mathrm{Hom}_A(L,A)$ and d is defined as above. More specifically (i) $d: \Lambda^0(L^*) \to \Lambda^1(L^*)$ is zero,

(ii) $d: \Lambda^1(L^*) \to \Lambda^2(L^*)$ is given by $d\omega(X,Y) = -\omega([X,Y])$,

(iii) d is extended to all $\Lambda^p(L^*)$ by requiring that it be a (signed) derivation

It is trivial to check that these three properties ensure $d^2 = 0$. Moreover, as $\Lambda_A^*(L^*)$ is an algebra over A, and d is an A-linear derivation, the cohomology $H^*(L)$ is a graded algebra over A. It is obvious that $L \to H^*(L)$ is a contravariant functor on the appropriate categories.

Another useful definition of d is in terms of the adjoint action of L on $\Lambda^*(L^*)$, and the interior product. Recall that if $X \in L$, $\omega \in \Lambda^p(L^*)$ then the interior product of ω with X, denoted $i_X\omega$ is by definition, $i_X\omega(X_1,\ldots,X_{p-1}) = \omega(X,X_1,\ldots X_{p-1})$. One checks that $i_X^2 = 0$, and that i_X is a derivation. Similarly, if $X \in L$, $\omega \in \Lambda^1(L^*) = L^*$, the adjoint action of X on ω is given by $L_X\omega(Y) = \omega([X,Y])$; and the operator L_X is extended to be an endomorphism of each $\Lambda^p(L^*)$ by requiring it to be a (signed) derivation. In geometry, where X represents a vector field, ω a form, L_X is the lie-derivative of ω with respect to X; and it is a basic formula in the subject that for any p-form, $L_X\omega = i_X d\omega + di_X\omega$. (Note that specifying $i_X d\omega$ for all X uniquely specifies $d\omega$.) This definition agrees with the previous one, as a simple check shows, because everything is a derivation.

2. The Chevalley-Eilenberg theorem

This theorem says that if G is a compact, connected lie group then there are canonical isomorphisms (i) $H^*(G:R) \tilde{\to} H^*(\underline{g})$, (ii) $H^*(\underline{g}) \simeq \Lambda^*(\underline{g}^*)^{\underline{g}}$ where $\underline{g}$ = lie algebra of G, and the superscript $\underline{g}$ refers to invariants under the adjoint action. We shall prove this result in all detail, because its usefulness goes far beyond compact groups, and because its proof is really much simpler than most texts make it out to be.

First recall "Maschke's theorem"; this says that given a short exact sequence $0 \to M_1 \to M \overset{\pi}{\to} M_2 \to 0$ of real or complex G-modules (G compact throughout this section), the sequence of invariant submodules $0 \to M_1^G \to M^G \xrightarrow{\pi^G} M_2^G \to 0$ is exact as well. The only assertion requiring proof is the surjectivity of π^G. (In fact the rest is true for any group whatsoever.) So choose a linear section $\sigma: M_2 \to M$ for π; using the normalized Haar integral for G (so that $\int_G dg = 1$) we average σ to get a map $\tau: M_2 \to M$ given by $\tau(m_2) = \int_G g\sigma(g^{-1} m_2)\, dg$. τ is now a G-equivariant section for π, hence $\tau^G: M_2^G \to M^G$ is a section for π^G, proving the surjectivity. It follows that if C^* is any cochain complex of real or complex G-modules (and G maps) then (C^{*G} being a subcomplex) $H^*(C^G) \simeq H^*(C)^G$, induced by the inclusion $C^{*G} \to C^*$.

The Chevalley-Eilenberg theorem follows from two applications of this last assertion. First consider the complex of all smooth forms $A^*(G)$. This is a complex of G-modules under left-translation, and $\Lambda^*(\underline{g}^*) = A^*(G)^G$. Thus $H^*(\underline{g}) \stackrel{\text{def}}{=} H^*(\Lambda(\underline{g}^*)) = H^*(A^*(G))^G \stackrel{\text{def}}{=} H^*_{DR}(G)^G$, where H_{DR} = deRham cohomology. But now de Rham cohomology is homotopy-invariant and G is path-connected, so $H^*_{DR}(G)^G = H^*_{DR}(G)$. This proves the first part of the theorem.

We now apply the same assertion to the complex $\Lambda^*(\underline{g}^*)$, regarded as a complex of G-modules under the adjoint action. It is trivial to see that d commutes with the G-action, because the adjoint action comes from a map of the underlying spaces. Thus we have $H^*(\Lambda(\underline{g}^*)^G) \simeq H^*(\Lambda(\underline{g}^*))^G$. We must now show that on the left-hand side, every G-invariant form is closed, and on the right-hand side that G acts trivially on the cohomology. For this purpose it is easier to deal with the adjoint action of $\underline{g}$, recalling that if M is a G-module, then $M^G = \{m \in M\colon Xm = 0$ for all $X \in \underline{g}\}$. (G is connected.)

First consider the right-hand side: the action of $\underline{g}$ on the complex $\Lambda^*(\underline{g}^*)$ is given by $L_X = di_X + i_X d$. This shows that for each X, the chain map L_X is chain-homotopic to zero, so induces the zero map on cohomology: that is, $H^*(\Lambda(\underline{g}^*))^{\underline{g}} = H^*(\Lambda(\underline{g}^*))$. For the left-hand side, we express d in terms of L_X as follows: choose a basis $X_1,\ldots,X_n \in \underline{g}$ and a dual basis $\xi_1,\ldots,\xi_n$ for $\underline{g}^*$. Then $\sum_{j=1}^{n} \xi_j \wedge L_{X_j} = (\sum_{j=1}^{n} \xi_j \wedge i_{X_j}) \circ d + \sum_{j=1}^{n} \xi_j \wedge diX_j$. It is trivial to check that $\sum_{j=1}^{n} \xi_j \wedge i_{X_j}$ = identity, and that the second summand is zero. Hence $d = \sum \xi_j \wedge L_{X_j}$, so if $\omega \in \Lambda^*(\underline{g}^*)^{\underline{g}}$, then $d\omega = 0$: thus the subcomplex $\Lambda^*(\underline{g}^*)^{\underline{g}}$ has zero differential and so $H^*(\Lambda(\underline{g}^*)^{\underline{g}}) = \Lambda^*(\underline{g}^*)^{\underline{g}}$.

The Chevalley-Eilenberg theorem is sometimes proved using Hodge's theory of harmonic forms; although this is a very pretty application of Hodge's theory, and shows that compact lie groups comprise a class of spaces on which the product of harmonic forms is harmonic, the above proof is really much simpler and more elementary.

<u>Computations</u>. We shall now show that the Chevalley-Eilenberg theorem, combined with Hopf's theorem, determines the cohomology of every reductive lie algebra over R or C. To begin with, let us just mention that the result checks for a torus: if T is an n-dimensional torus with Lie algebra t, then clearly $H^*(T:R) = \Lambda_R^*(t^*)$; since t is abelian the differential in $\Lambda_R^*(t^*)$ is zero, so $H^*(t) = \Lambda_R^*(t^*)$; and for the same reason $\Lambda^*(t^*)^t = \Lambda^*(t^*)$.

1. Consider $\underline{gl}(n,R)$. This is not the Lie algebra of a compact group. However, $\underline{gl}(n,R) \otimes C = \underline{gl}(n,C)$ and if u(n) denotes the lie algebra of the unitary group (which is compact) then $u(n) \otimes C = \underline{gl}(n,C)$ as well. Hence $H^*(u(n)) \otimes C = H^*(u(n) \otimes C) = H^*(\underline{gl}(n,C)) = H^*(\underline{gl}(n,R)) \otimes C$. Now $H^*(u(n)) = H^*(U(n))$, and we know that this last object is an exterior algebra on generators of degree 1,3,...,2n-1. Thus the complexified algebra $H^*(\underline{gl}(n,R)) \otimes C$ is exterior as well (on generators of degree 1,3,...). From the universality of exterior algebras it

follows that if A^* is a real algebra whose complexification† is exterior, then A^* is exterior on generators of the same degree. Hence $H^*(\underline{gl}(n,R)) = \Lambda^*(x_1, x_3, \ldots, x_{2n-1})$.

2. Quite generally, if $\underline{g}$ is a real reductive lie algebra, there is a lie algebra k of a compact group K such that $\underline{g} \otimes C = k \otimes C$. Since we know $H^*(k) = H^*(K)$, and that this latter algebra is exterior, it follows exactly as above that $H^*(\underline{g})$ is an exterior algebra, on generators of the same degree as those for K.

3. From number 2 above we see that if G^c is a connected reductive complex group, then $H^*(G^c:C) = H^*(\underline{g}^c:C)$; this is because the compact group K in question is the maximal compact subgroup of G^c, so G^c has the same homotopy-type as K (by polar decomposition). For <u>real</u> reductive groups this isomorphism does <u>not</u> hold. A simple useful example to keep in mind is this. $H^*(SU(2):C) = H^*(SL_2(C):C) = H^*(sl(2,C))$: however, $H^*(SL_2(R)) = \Lambda(x_1)$, $\deg x_1 = 1$ since $SL_2(R)$ is homotopy-equivalent to $SO(2) = S^1$, whereas $H^*(sl(2,R)) = \Lambda(x_3)$, $\deg x_3 = 3$ as one sees either from the general assertion above (No. 2) or by simple direct verification.

This distinction will play an important part in the computations of the exotic characteristic classes for foliations on homogeneous spaces in Chapter II. The point is that at the lie algebra level, many of the computations will be parallel for the real and complex case: however, to get topologically

†More generally, this works for any ring extension.

significant answers, in the real case one will often have to pass to a quotient by a suitable discrete group.

4. Finally, the assertion of No. 2 above shows that if $\underline{g}$ is a reductive lie algebra, then $H^*(\underline{g})$ satisfies a "Poincaré-duality". For a purely algebraic proof, see Koszul [16].

3. Relative cohomology

In analogy with the first section, we shall now look at the cohomology of invariant forms on a homogeneous space in order to arrive at a purely algebraic definition of the cohomology of a lie algebra, relative to a subalgebra. So consider a lie group G, a closed subgroup $H \subset G$ and the homogeneous space G/H; G acts on G/H from the left, and exactly as in Section 1, the left G-invariant differential forms comprise a subcomplex of the complex of all differential forms on G/H. This subcomplex is easier to describe if we pull back the cotangent bundle over G.

If $\underline{g}$ and $\underline{h}$ are the lie algebras of G, H respectively, then the cotangent bundle $T^*(G/H)$ pulled back over G is canonically trivialized as $G \times (\underline{g}/\underline{h})^*$, where $(\underline{g}/\underline{h})^* = \mathrm{Hom}_R(\underline{g}/\underline{h}, R)$. Thus sections of exterior powers of this bundle over G form a subalgebra of sections of $G \times \Lambda^*(\underline{g}^*)$, which is $A^*(G)$. Using the right action of H on G, and the adjoint action of H on $\Lambda^*(\underline{g}/\underline{h})^*)$, one easily checks that the G-invariant forms on G/H correspond to $\Lambda^*(\underline{g}/\underline{h})^*)^{\underline{h}} \subseteq \Lambda^*(\underline{g}^*)$. Moreover, using the relation between d, L_X and i_X one sees that $\Lambda^*((\underline{g}/\underline{h})^*)^{\underline{h}}$ is a sub-DGA of $\Lambda^*(\underline{g}^*)$. This model gives our algebraic definition.

Definition. Let L be a lie algebra, L' a subalgebra (over an arbitrary ring A). The relative cohomology $H^*(L,L^*)$ is by definition the cohomology of the complex $(\Lambda((L/L')^*)^{L'},d)$ where d is restricted from $\Lambda^*(L^*)$.

When G is compact (and thus also H), the Chevalley-Eilenberg theorem identifies $H^*(\underline{g},\underline{h})$ with the (de Rham) cohomology $H^*(G/H)$. The proof is essentially a repetition of the first part of Section 2.

4. Cohomology with coefficients

Let L be a Lie algebra over a ring, M and L-module. We will define the cohomology of L with coefficients in M. In a very special case, this gives the de Rham cohomology of a group (see below) although the better analogy in this case is with the Eilenberg-MacLane cohomology.

Suppose we consider a lie group G with lie algebra $\underline{g}$. G acts by translations on the space $C^\infty(G)$ of smooth functions, and hence so does $\underline{g}$ (by the usual vector-field operation). Reversing our point of view from the first section, one can regard the algebra of all smooth forms $A^*(G)$ as $\Lambda^*(\underline{g}^*) \otimes_R C^\infty(G)$, and ask for a description of d in this big complex in terms of the d in the invariant complex, where we now regard the latter d as a God-given algebraic object. Consider d_A on $A^0(G)$ for example; thus given a smooth function $f \in C^\infty(G) = A^0(G)$, $d_A(f)(X) = X(f)$. Next consider a smooth one-form $\omega \in A'(G)$: $d_A\omega(X,Y) = X\omega(Y) - Y\omega(X) - \omega(X,Y)$. In this last formula, if we write $\omega = \sum \omega_i \otimes f_i$

where $\omega_i \in \Lambda^1(\underline{g}^*)$, $f_i \in C^\infty(G)$ and take $X,Y \in \underline{g}$ we see at once that d_A can be defined by $i_X d_A \omega(Y) = L_X\omega(Y) - d_A i_X \omega(Y)$. This gives the clue then to defining the algebraic cohomology.

Definition. The cohomology modules $H^*(L:M)$ are the cohomology of the complex $(\Lambda^*(L^*) \otimes_R M, d_M)$ where d_M is defined inductively, as follows: $d_M^0: M \to \Lambda^1(L^*) \otimes M$ is given by $i_X d_M^0(m) = x \circ m$ and d_M^p is then given by $i_X d_M^p \omega = x \circ \omega - d_M^{p-1}\ i_X \omega$, where $x \circ \omega$ denotes the action of L on the product module $\Lambda^p(L^*) \otimes M$. Explicitly, d_M^p can be written out as follows.

$$d_M^p \omega(x_0,\ldots,x_p) = \sum_{i=0}^{p} x_i \cdot \omega(x_0,\ldots,x_i,\ldots,x_p)$$

$$+ \sum (-1)^{i+j}\ \omega([x_i,x_j]\ x_0,\ldots\hat{x}_i,\ldots\hat{x}_j,\ldots x_p)$$

As the reader may have suspected, one can equally well define relative cohomology with coefficients in a module, $H^*(L,L':M)$. This is the cohomology of the complex $(\Lambda^*((L/L')^*) \otimes M)^{L'}$ $\subseteq \Lambda^*(L^*) \otimes M$, with the differential inherited from the larger complex.

5. The spectral sequence

Given a lie algebra L, and L-module M and a subalgebra L', there is a spectral sequence whose E_1-term is $H^*(L': \Lambda^*(L,L':M)$ and which converges to $H^*(L:M)$. If L' is not an ideal, there is no convenient description of the E_2-term in general. The details can be found in the paper of Hoschschild and Serre [13]: here we simply give the filtration.

We filter the complex $\Lambda^*(L^*) \otimes M$ (relative to L') as follows: for $p \leq q$

$$F^p_{L'}(\Lambda^q(L^*) \otimes M) = \left\{\omega \in \Lambda^q(L^*) \otimes M: \omega(x_1,\ldots,x_q) = 0 \right.$$
$$\left. \text{whenever} \quad x_1,\ldots,x_{p+1} \in L'\right\}.$$

(If the characteristic of the base ring is not 2-divisible, this is all right: otherwise one has to define $\omega(x_1,\ldots,x_q) = 0$ whenever any $q - p + 1$ components are in L').

The E_0-term is then $\Lambda^*(L'^*) \otimes (\Lambda^*((L/L')^*) \otimes M)$ regarded as the complex for the cohomology of L' with coefficients in the graded L'-module $\Lambda^*(L/L')^* \otimes M$, and this gives the E_1-term above.

References

1 A. Borel, Compact Clifford-Klein forms of symmetric spaces. Topology 2 (1963) 111-122.

2 R. Bott, Lectures on characteristic classes and foliations. Lectures on Algebraic and Differential Topology. R. Bott et al. ed. Lecture notes in mathematics 279, 1-94. (New York: Springer-Verlag, 1972)

3 R. Bott and A. Haefliger, On characteristic classes of Γ-foliations. Amer. Math. Soc. Bull. 78 (1972) 1039-1044.

4 R. Bott and J. Heitsch, A remark on the integral cohomology of BΓq. Topology 11 (1972) 141-146.

5 H. Cartan, Notions d'algebra differentielle,... CBRM Colloque de Topologie, Brussels 1950 15-27. (Liège: G. Thome, 1951)

__________, La transgression dans un groupe de Lie et dans un espace fibré principal. CBRM Colloque de Topologie, Brussels 1950 57-71. (Liege: G. Thone, 1951)

6 S. Chern, Characteristic classes of Hermitian manifolds. Annals of Math. 47 (1946) 85-121.

7 D. B. Fuks, Characteristic classes of foliations. Russian Math. Surveys 28 (1973), No. 2, 1-16.

8 I. M. Gelfand, B. L. Feigin and D. B. Fuks, Cohomologies of the Lie algebras of formal vector fields with coefficients in its adjoint space and variations of characteristic classes of foliations. Funct. Anal. and its Appl. 8 (1974) 99-112.

9 C. Godbillon, Cohomologies d'algèbres de Lie de champs de vecteurs formels. Sem, Bourbaki 1972/73. Exposé 421. Lecture Notes in mathematics 383 69-87. (New York: Springer-Verlag, 1974)

10 C. Godbillon and J. Vey, Un invariant des feuilletages de codimension 1. Acad. des Sciences. Paris. Comptes Rendus. ser. A. 273 (1971) 92-95.

11 A. Haefliger, Sur les classes charactéristiques des feuilletages. Sem. Bourbaki 1971/72. Exposé 412. Lecture notes in mathematics 317. 239-260. (New York: Springer-Verlag, 1973)

12 J. L. Heitsch, Deformation of secondary characteristic classes. Topology 12 (1973) 381-388.

13 G. Hochschild and J-P. Serre, Cohomology of Lie algebras. Annals of Math. 57 (1953) 591-603.

14 S. Kobayashi, Canonical forms on frame bundles of higher order contact. Symposia in pure math. 3 (1960) 186-193. (Providence: Amer. Math. Soc. 1961)

15 S. Kobayashi and K. Nomizu, Foundations of differential geometry. Vol. I. (New York: Interscience-Wiley, 1963)

16 J-L. Koszul, Homologie et cohomologie des algèbres de Lie. Soc. Math. France, Bull. 78 (1950) 65-127.

17 J. W. Milnor and J. D. Stasheff, Characteristic classes. Annals of Math. Studies, 76. (Princeton: Princeton Univ. Press, 1974)

18 H. Shulman, Secondary obstructions to foliations. Topology 13 (1974) 177-183.

19 J. Simons, Characteristic forms and trangression II. Annals of Math. (to appear).

20 W. Thurston, The theory of foliations of codimension greater than one. Commentarii Math. Helv. 49 (1974) 214-231.

21 W. Thurston, Noncobordant foliations of S^3. Amer. Math. Soc. Bull. 78 (1972) 511-514.

22 J. A. Wolf, Spaces of constant curvature. 2nd ed. (Berkeley: Joseph A. Wolf, 1972)